普通高等院校"十三五"规划教材

现代科技创新成果英文赏析

Appreciation of English Passages of Modern Scientific and Technological Innovation Achievements

王凤琴　康　勇　马　静　编著

西南交通大学出版社
·成　都·

图书在版编目（CIP）数据

现代科技创新成果英文赏析 / 王凤琴，康勇，马静编著. —成都：西南交通大学出版社，2019.4
普通高等院校"十三五"规划教材
ISBN 978-7-5643-6817-3

Ⅰ. ①现⋯ Ⅱ. ①王⋯ ②康⋯ ③马⋯ Ⅲ. ①科技成果–世界–英语–高等学校–教材 Ⅳ. ①N11

中国版本图书馆 CIP 数据核字（2019）第 060197 号

普通高等院校"十三五"规划教材

现代科技创新成果英文赏析
Xiandai Keji Chuangxin Chengguo Yingwen Shangxi

王凤琴　康　勇　马　静　**编著**

责 任 编 辑	孟　媛
封 面 设 计	原谋书装
出 版 发 行	西南交通大学出版社 （四川省成都市二环路北一段 111 号 西南交通大学创新大厦 21 楼）
发行部电话	028-87600564　028-87600533
邮 政 编 码	610031
网　　　址	http://www.xnjdcbs.com
印　　　刷	四川森林印务有限责任公司
成 品 尺 寸	185 mm × 260 mm
印　　　张	11.25
字　　　数	310 千
版　　　次	2019 年 4 月第 1 版
印　　　次	2019 年 4 月第 1 次
书　　　号	ISBN 978-7-5643-6817-3
定　　　价	32.00 元

课件咨询电话：028-87600533
图书如有印装质量问题　本社负责退换
版权所有　盗版必究　举报电话：028-87600562

前 言
PREFACE

在当今这个科技知识爆炸的时代，不断出现的创新成果就如一个个原动机，推动着人类社会文明发展的列车向着更高更远的目标快速前进。了解和掌握国外最新的科技创新成果，不断追踪世界科技发展的最前端，为赶超世界科技强国提供更多更新的科技信息，对于广大立志不断创新的读者十分重要。本书是一本集科技成果为一体的教材，读者不但能直接阅读和学习最新的科技英文短文，方便快速地掌握最新的科技成果，而且还能学习涉及这些成果的英文词汇及表达方式。本书不但拓宽读者的视野，还能激发出读者的智慧火花，产生新意识、新思想、新概念，找到新方法和新手段。

本书精选了五十七篇现代科技英语短文，通过阅读本书，从中读者可以学到地道的科技英语最新表达方法，追踪当代科学技术发展前缘及了解到人类取得的成果。五十七篇短文主题包括目前人们普遍关心的与人文、社会及自然科学等学科相互渗透的新型学科的发展。这些短文对拓宽读者文化视野，培养其创新意识，使其掌握现代科技发展的前沿动态，激发其对创新的兴趣和创新意识有着积极的意义。

本书由王凤琴教授、康勇教授及马静老师编写。本书的出版得到了西安石油大学出版基金的资助及相关专业人士的大力支持，编者在此对所选文章的作者一并表示衷心的感谢。

由于时间仓促、编者水平有限，书中难免存在疏漏之处，敬请读者批评指出。

编 者
2019 年 3 月

目 录
CONTENTS

Passage 1	Ants Farm Plantsi	1
Passage 2	New Alloy	4
Passage 3	Positive Effect of Carbon Dioxide Levels	7
Passage 4	Robotic Cat for the Elderly	10
Passage 5	Insects	13
Passage 6	Story of Petroleum	16
Passage 7	Artificial Life	19
Passage 8	3D Food Printer	22
Passage 9	Next Innovations	25
Passage 10	Globe-spanning Telescope	28
Passage 11	Bionic Leaf	31
Passage 12	Brain Damage	34
Passage 13	Changing Smart Phone into a Microscope	37
Passage 14	Disabled People Can Move Again	40
Passage 15	First Quantum Satellite	43
Passage 16	Lie Detector App	46
Passage 17	A New Skin Sensor	49
Passage 18	Quinine	52
Passage 19	Renewable Biofuels from Microalgae	55
Passage 20	Solar Water Disinfection	58
Passage 21	Solar Energy	61
Passage 22	Happy Bees	64
Passage 23	Newly Discovered State of Memory	67
Passage 24	Advances in Intelligent Vehicles	70
Passage 25	Compulsive Behavior	73
Passage 26	Manned Spaceship	76
Passage 27	Analysis and Construction of Plastic Degradation	80
Passage 28	Atlantic Salmon	83

Passage 29	Blue Halos	86
Passage 30	Childhood Tumors	89
Passage 31	Do Animals Recognize Themselves?	92
Passage 32	Glass Could Hold Ancient Fossils	95
Passage 33	Laughter	98
Passage 34	Explores the Martian Upper Atmosphere	101
Passage 35	Overweight and Obesity	105
Passage 36	About Soil	108
Passage 37	Solids and Liquids	111
Passage 38	World's Largest Solar Furnace	114
Passage 39	Self-healing Polymeric Material	117
Passage 40	Unified Theory of Evolution	120
Passage 41	Creativity and Childhood Experiences	123
Passage 42	The Risk of Excessive Dietary Salt	126
Passage 43	The Discovery of Artemisinin	129
Passage 44	NASA's Plans to Send Humans to Mars	132
Passage 45	New Artificial Nerves	135
Passage 46	Blood Donation and Cardiovascular Disease	138
Passage 47	Superglue	141
Passage 48	The Pavements that Generate Solar Energy	144
Passage 49	The Role of Embryonic Stem Cells	147
Passage 50	The Secret in the Fingerprint	150
Passage 51	Bees Have Different Taste in Flowers	153
Passage 52	Ants and Free Radicals	156
Passage 53	Sleeping	159
Passage 54	An Anti-aging Method	162
Passage 55	Smog Can Harm Your Heart	165
Passage 56	Two Myths of Snow	168
Passage 57	Unmanned Aerial Vehicle	171
References		174

Passage 1 Ants Farm Plantsi

A species of ant in Fiji survives by farming some plant species. The ants nurture seedlings and then live in the cavities of the plants.

The researches found that these ants carry seeds of six species of plant and insert them into cracks in trees, where they germinate. The ants also fertilize the seeds with their waste, and are only found living near these plants. Although ants are known to farm fungi, this is the first time they have been found to plant seeds and actively cultivate them. Examining the family trees of relatives of these ants and plants suggests that this mutually dependent relationship evolved around 3 million years ago.

Word bank

1. a species of 一种
2. nurture ['nɜːtʃə(r)] *v.* 培育；养育 *n.* 教养；培育
3. cavity ['kævəti] *n.* 腔，洞；蛀牙，龋洞 oral cavity 口腔；nasal cavity 鼻腔
4. crack [kræk] *n.* 裂缝；缝隙 *v.* 破裂；打开
5. germinate ['dʒɜːmɪneɪt] *v.* 发芽；生长；发育
6. fungi ['fʌŋgiː] *n.* (fungus 的复数)真菌；霉菌
7. cultivate ['kʌltɪveɪt] *v.* 耕作；种植；培养
8. mutually ['mjuːtʃuəli] *adv.* 相互地；彼此；共同地
9. evolve [ɪ'vɒlv] *v.* 发展；进化

Exercises

I. Fill in the blanks with the words given below. Change the form where necessary.

germinate	crack	fungi	evolve
cavity	elaborate	nurture	mutually

1. Let's find a _____ convenient time to meet.
2. This is a good habit to _____.
3. Popular music _____ from folk songs.
4. There's a _____ in the boy's tooth.
5. Seeds will not _____ without water.
6. The plate had a _____ in it.

II. Comprehension of the passage.

Choose the best answer to each of the following questions.

1. Where do the ants live? _____
   A. They live in their cavities.
   B. They live in the cavities of the plant.
   C. They live near the species of the plants.
   D. Both B and C.

2. The paragraph mainly talks about _____.
   A. the ant in Fiji survives by farming six plant species
   B. the relation between the ants and the six plants
   C. a species of ants can farm the plants
   D. the plants are important to the ants' lives

3. Which of the following is wrong about this passage? _____
   A. The ants carry seeds of six plants and insert them into the cracks in the tree.
   B. It isn't the first time for the researchers to find the seeds and cultivate them.
   C. Not every ant has capacity to farm and cultivate the plants.
   D. Mutually dependent relationship between ants and plants evolved several million years ago.

4. The author presents details according to _____.
   A. simple listing   B. statistics
   C. time order   D. spatial order

Reference translation

会种植物的蚂蚁

斐济有一种蚂蚁，它们会种植某些植物来养活自己。这些蚂蚁把植物的种子种在所依附的植物上，然后住在植物的空洞内。

研究人员发现，这些蚂蚁会把六种植物种子种入树上的缝隙里，让其在那里生根发芽。蚂蚁还会用自己的排泄物来对其施肥。这些蚂蚁就可以依赖这些植物作为栖身之地。虽然人们知道蚂蚁会培育真菌，但这是第一次发现蚂蚁会利用植物种子努力地培育植物。通过研究这种蚂蚁和这些植物的依存关系，人们发现，它们之间相互依赖的关系至今已有大约三百万年了。

Passage 2 New Alloy

Memory alloys that spring back into a pre-defined shape are nothing new, but regular bending means they will fatigue and fail within a relatively short time-scale. Now, a team of engineers has developed an alloy that rebounds into shape even after 10 million bends. The new alloy, which is made from nickel, titanium and copper, has a special crystal structure that allows it to undergo bending more easily than most metals. The constituent atoms are arranged in a way that allows them to switch between two different configurations, over and over. This is known as a phase transition, and it can occur either with changes in temperature or merely the release of tension.

The result is published in *Science*. The researchers suggest that the new material could be used in situations where loading varies constantly, like in the wings of airplanes, or where materials undergo regular heating and cooling.

Word bank

1. alloy ['ælˌɔɪ] *n.* 合金；*v.* 合铸，熔合（金属）
2. spring [sprɪŋ] *v.* 弹；跳 *n.* 弹簧；弹性；春天
3. rebound [rɪ'baʊnd] *v.* 弹回
4. titanium [tɪ'teɪniəm] *n.* [化]钛
5. undergo [ˌʌndə'gəʊ] *v.* (underwent, undergone) 经历；遭受；承受
6. constituent [kən'stɪtjʊənt] *n.* 成分；构成要素 *adj.* 构成的；组成的
7. switch [swɪtʃ] *v.* 转换；转变；交换
8. configuration [kənˌfɪgə'reɪʃn] *n.* （分子中原子的）组态，排列；构造；布局
9. release [rɪ'liːs] *n.* 释放；发布

Exercises

I. Fill in the blanks with the words given below. Change the form where necessary.

undergo	constituent	spring	release
rebound	configuration	turbidity	switch

1. Plants _____ water through their leaves by transpiration.
2. My mother _____ major surgery last year.
3. The ball _____ from the goalpost.
4. Sugar is the main _____ of candy.
5. The dates of the last two exams have been _____.
6. The branch _____ back and hit him in the face.

II. Comprehension of the passage.

Choose the best answer to each of the following questions.

1. Compared with the old alloy, the new alloy _____.
   A. can rebound into shape even after 10 million
   B. can be used widely
   C. is produced easily
   D. is expensive
2. The new alloy is made from _____.
   A. nickel   B. titanium
   C. copper   D. nickel, titanium and copper
3. The new alloy can be used in _____.
   A. airplane   B. pigboat
   C. engine   D. rocket
4. The topic of this passage is _____.
   A. a brief introduction of the new alloy
   B. an application of the new alloy
   C. the prospect of the new alloy
   D. the function of the new alloy

Reference translation

新合金

  能够回弹到初始形状的记忆合金并不是什么新鲜事，但是长期弯曲意味着它们会在相对比较短的时间范围内疲劳并失效。现在，有一工程师团队开发出了一种合金，即使经过一千万次弯曲后，还能恢复原形。这种由镍、钛和铜制成的新合金具有特殊的晶体结构，比大多数金属更容易弯曲。其原子的排列方式允许它在两个不同的排列之间切换。这就是所谓的相变，它可以发生在温度变化时或仅仅在释放张力时。

  该成果在发表在《科学》杂志上。研究人员建议，这种新材料可用在不断承受压力改变的地方，比如机翼或者规律性冷热变化的地方。

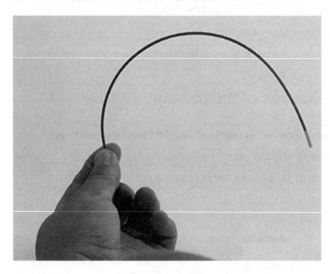

Passage 3 Positive Effect of Carbon Dioxide Levels

A new study suggests that carbon dioxide levels caused by human activity had a positive effect.

They have led to a massive increase in plant growth. Using satellite data, experts calculate that between a quarter and a half of the earth's vegetated land has become greener over the thirty-three years since the records began. The scientists say the rising global temperature has helped fuel the plant boom along with more nitrogen in environment and shifts in land management. But they estimate that seventy percent of the change has been caused by plants absorbing more of our emissions of carbon dioxide because CO_2 is a plant fertilizer. This is good news because as plants soak up the carbon, they are slowing somewhat the pace of climate change by keeping the carbon out of the atmosphere.

Word bank

1. massive ['mæsɪv] *adj.* 大规模的；大量的
2. calculate ['kælkjʊleɪt] *v.* 计算；估计；打算
3. vegetate ['vedʒəteɪt] *v.* 生长；过单调呆板的生活
4. nitrogen ['naɪtrədʒən] *n.* [化]氮，氮气
5. emission [ɪ'mɪʃn] *n.* 排放；辐射；排放物
6. fertilizer ['fɜːtəlaɪzə(r)] *n.* 肥料；化肥
7. soak [səʊk] *v.* 浸泡；吸入 soak up 吸收
8. somewhat ['sʌmwɒt] *adv.* 有点；有几分；稍微

Exercises

I. Fill in the blanks with the words given below. Change the form where necessary.

| vegetate | nitrogen | soak up | emission |
| somewhat | fertilizer | amnesia | massive |

1. The organic _____ shall keep the soil in good heart.
2. The plant requires _____ in order to make proteins.
3. The fields _____ vigorously.
4. I was _____ surprised to see him.
5. The explosion made a _____ hole in the ground.
6. The cells will instantly start to _____ moisture.

II. Comprehension of the passage.

Choose the best answer to each of the following questions.
1. Carbon dioxide is positive in _____.
   A. helping the plant to grow
   B. shifts in land management
   C. fruit and vegetable insurance
   D. production of carbonated beverage combination
2. Carbon dioxide has effect on the pace of climate change by_____.
   A. thick insulation layer
   B. keeping the carbon out of the atmosphere as plants soak up the carbon
   C. preventing sunlight from entering
   D. absorbing a large amount of heat
3. Carbon dioxide may help plants growth because of_____.
   A. more nitrogen in environment and shifts in land management
   B. the slow pace of climate change
   C. plants absorbing more of emissions of carbon dioxide
   D. both A and C
4. The author's attitude towards the carbon dioxide is _____.
   A. indifferent   B. biased
   C. negative   D. positive

Reference translation

二氧化碳的积极作用

一项新的研究表明,由人类活动引起的二氧化碳具有积极的作用。

它们会大大促进植物生长。专家们根据卫星提供的数据计算出了地球上四分之一到一半植物覆盖的土地面积比30年前开始记录以来有所增加。科学家称,全球气温逐渐升高,同时环境中氮气也在增加,土地转化加快,有助于供给植物生长。但他们估计,其中70%的变化是由植物吸收了更多的二氧化碳引起的,因为二氧化碳是植物的肥料。这是个好消息,因为植物吸收二氧化碳,减少了大气中的二氧化碳,从而在某种程度上也降低了气候变化的速度。

Passage 4 Robotic Cat for the Elderly

"He feels like a real cat," says Jim McGuckin as I place Alan on his lap. And Alan is the name I've given to a robotic "companion pet", developed by toy maker Hasbro. Research suggests there is a real benefit to providing people with companion robots, particularly if they are suffering from dementia or Alzheimer's disease.

Paro is a Japanese-designed robotic seal. It's without question the cutest thing in the room when it's on a tech show, and great thought has gone into making Paro is a delight—you plug him in by putting a dummy into his mouth. It's significantly more advanced than Hasbro's cat, but also significantly more expensive. Paro is touted as a medical device. Paro is in use around the world and will look at expanding the number of Paros in hospitals and care homes.

Word bank

1. range [reɪndʒ] *n.* 范围；类别
2. companion [kəm'pænɪən] *n.* 陪伴；伙伴；同伴
3. combat ['kɒmbæt] *v.* 与……战斗；与……斗争；*n.* 战斗；格斗
4. initiative [ɪ'nɪʃətɪv] *n.* 倡议；新方案；主动性；主动权
5. dementia [dɪ'menʃə] *n.* [医]痴呆
6. Alzheimer ['ælz'ɛmə] *n.* 老年痴呆症；阿尔茨海默病（亦称 Alzheimer's disease）
7. dummy ['dʌmi] *n.* 奶嘴；仿制品；笨蛋
8. plug [plʌg] *n.* 插头；塞子 *v.* 插入
   plug in 插上电源的插头；堵住……上的洞
9. tout [taʊt] *v.* 兜售；推销；标榜；吹捧

Exercises

I. Fill in the blanks with the words given below. Change the form where necessary.

| combat | plug | tout | dementia |
| companion | fertilizer | range | initiative |

1. _____ is the name for the effects of Alzheimer's disease, stroke and other brain disorders.

2. There is a full _____ of activities for children.

3. We become _____ in misfortune from then on.

4. She's being _____ as the next leader of the party.

5. She did it on her own _____.

6. He is determined to _____ with his bad habits.

II. Comprehension of the passage.

Choose the best answer to each of the following questions.

1. The robotic cat is designed to _____.
    A. play with kids
    B. help the disable
    C. help homeless people
    D. accompany the loneliness elderly

2. _____ needs the robotic cat most.
    A. An old man was suffering from dementia
    B. An old man having a trip with friends
    C. An old man loves animals
    D. An old man wants to be healthy

3. Which of the following is NOT true? _____
    A. Paro was designed by Japanese.
    B. Paro needs to recharge.
    C. Paro's plug looks like dummy.
    D. Paro is an old cat who likes play with old people.

4. From the passage, we know that _____.
    A. Hasbro designed Paro

B. robotic cat was used by more old people
C. robotic cat was cheaper
D. every old people needs robotic cat

Reference translation

陪伴老人的机器猫

"它摸起来像一只真的猫。"当我把艾伦放在吉姆·麦格金的膝盖上时,他这样说道。艾伦是玩具制造商孩之宝公司开发的"陪伴宠物"机器人。研究表明,陪伴机器人的发明确实给人们带来了益处,尤其是对那些患有痴呆症和阿尔茨海默病的人。

帕罗是一款由日本设计的海豹机器人,当它在科技展上露面的时候,毫无疑问是最可爱的。它的充电器被设计成一个奶嘴的模样,充电时把它放进帕罗的口中就可以了。很明显帕罗比孩之宝玩具公司的猫更先进,但也更贵。帕罗被定义为一种医疗设备,它将在全世界范围内推广使用,更多地应用于医院和养老院。

Passage 5 Insects

Insects have spread to every possible place on the face of the globe. Everywhere man has gone, he has found insects there before him. Seven hundred miles north of the Arctic Circle, survivors were amazed to see a butterfly. Thirty-five thousand feet above sea level, a balloonist noticed a honeybee flying around the basket of his craft. Hundreds of miles from the nearest coast, mariners have seen water bugs skating over the surface of the swells.

High in the Himalayas, mountain climbers have found insects. A praying mantis was discovered 16,000 feet above sea level. In Ecuador butterflies were sighted among the crags of the Andes at 18,500 feet. A moth in South America is encountered all the way from lowland swamps up to 10,000 feet on mountainsides. At a height of 200 feet, scientists have figured, there is an insect for every 6,748 cubic feet of air. One mile above the earth, the ratio is one insect for each 117,546 cubic feet.

Word Bank

1. globe [gləʊb] *n.* 球；地球
2. the Arctic Circle *n.* 北极圈
3. survivor [sə'vaɪvə(r)] *n.* 生还者；幸存者
4. mariner ['mærɪnə(r)] *n.* 海员，水手
5. water bug *n.* 半翅类水虫，蟑螂
6. the Himalayas *n.* 喜马拉雅山脉
7. praying mantis *n.* 螳螂
8. crag [kræg] *n.* 峭壁
9. moth [mɒθ] *n.* 蛾，飞蛾
10. ratio ['reɪʃɪəʊ] *n.* (*pl.* ratios) 比；比率

Exercises

I. Fill in the blanks with the words given below. Change the form where necessary.

| intercept | duplicate | encode | entanglement |
| nickname | credited | decipher | quantum |

1. He climbed one _____ after another.
2. The _____ of boys and girls in this class is one to three.
3. She was the only _____ of the air crash.
4. 70% of our _____'s surface is water.
5. The _____ air-craft later crashed into the hillside.
6. A _____ must have his eye on rocks.

II. Comprehension of the passage.

Choose the best answer to each of the following questions or answer the question.

1. From this article we can see that _____.
   A. insects are resistant to low temperatures
   B. jungles contain more insects than deserts
   C. insects breed faster over water
   D. insects avoid mountainous areas

2. As one travels higher into the atmosphere, _____.
   A. he encounters strange looking insects
   B. he sees only tropical insects
   C. he finds helpless insects at the mercy of the winds (被风刮来刮去)
   D. he notices insects becoming fewer and fewer

3. The author arranges details according to _____.
   A. simple listing
   B. time order
   C. cause and effect
   D. contrasts

4. Underline the sentence which suggests that insects may be found at the South Pole (南极).

Reference translation

昆 虫

  昆虫已经蔓延到地球表面每一个可能的地方。人可以去的地方昆虫都先于到达。北极圈以北 700 英里①处，幸存者惊奇地看到一只蝴蝶。在海拔 35 万英尺②高空，一个热气球驾驶者发现一只蜜蜂在飞行器的吊篮周围飞行。在距海岸几百英里的海面上，海员们可以观察到汹涌的海面上水虫在滑行。

  在喜马拉雅山的高处，登山者发现了昆虫。一只螳螂出现在海拔 16 000 英尺处。在厄瓜多尔的安第斯山脉 18 500 英尺的峭壁上，人们发现了蝴蝶。在南美洲，从低地沼泽到高达 10 000 英尺的山地，都能看到飞蛾。科学家估计，在 200 英尺高的地方，每 6 748 立方英尺的空间就有一只昆虫。高于地球表面一英里处，昆虫存在率是每 117 546 立方英尺有一只。

---

① 1 英里 ≈ 1.6 千米。
② 1 英尺 ≈ 0.3 米。

Passage 6　Story of Petroleum

The story of petroleum probably goes back before the dawn of recorded history. In some places, where the rock layers of the earth's crust were broken, petroleum appeared at the surface of the ground. Early man very likely found this and used petroleum pitch or asphalt long before we have written records. We do know from some of the world's oldest writings that asphalt was used as mortar in buildings. We know that pitch was use in calking boats, and that ancient Egyptians used petroleum to grease their chariot axles. In America the native Indians used petroleum for fuel long before the white man came. The early settlers considered oil a nuisance because it interfered with getting salt brine from wells that bad been drilled for that purpose. Actually, the great petroleum industry got its start from bottled medicine oil and the fact that the whale oil used for lighting and lubrication was becoming very scarce.

Word Bank

1. petroleum [pə'trəuliəm] *n.* 石油
2. dawn [dɔ:n] *n.* 开端；黎明

3. crust [krʌst] *n.* 地壳；外壳；（糕点的）酥皮

4. pitch [pɪtʃ] *n.* [U] 沥青；柏油

5. asphalt ['æsfælt] *n.* 沥青；柏油

6. mortar ['mɔːtə(r)] *n.* 灰泥；砂浆

7. calk [kɔːk] *v.* 堵（船的）缝；使不漏水；填塞

8. grease [griːs] *v.* 涂油脂于，用油脂润滑 *n.* 动物油脂；润滑油

9. chariot ['tʃærɪət] *n.* （古代用于战争或竞赛的）敞篷双轮马车；战车

10. axle ['æksl] *n.* 车轴；轮轴

11. nuisance ['njuːsns] *n.* 讨厌的东西（人，行为）；麻烦事

12. interfere [ˌɪntə'fɪə(r)] *v.* 干预，干涉；妨碍 interfere with

13. brine [braɪn] *n.* 盐水 *v.* 用浓盐水处理（或浸泡）

14. scarce [skeəs] *adj.* 缺乏的；稀少的

Exercises

I. Fill in the blanks with the words given below. Change the form where necessary.

| grease | crust | interfere | nuisance |
| scarce | brine | dawn | pitch |

1. This will not _____ with my work. I promise.

2. It's a _____ having to go back tomorrow.

3. They start work at _____.

4. I _____ the wheels and adjusted the brakes.

5. Jobs are becoming increasingly _____.

6. Bake the bread until the _____ is golden.

II. Comprehension of the passage.

Choose the best answer to each of the following questions or answer the question.

1. Early man probably found petroleum as a result of _____.

   A. accidental discovery while hunting

   B. experimentation with natural substances

   C. following the advice of his elders

   D. reading what scholars had written

2. It is likely that petroleum was first used for _____.
    A. lubricating moving parts
    B. curing illnesses
    C. cooking food
    D. making structures water-tight
3. The early colonists _____.
    A. used oil to extract salt from water
    B. rejected the Indian's use of oil
    C. probably used oil to warm their homes
    D. drilled the first oil wells in America
4. Underline an expression which suggests that knowledge about the earliest uses of petroleum are purely speculative (猜测的).

Reference translation

石油的故事

石油的故事可能要追溯到史前。在一些地壳的岩石层开裂的地方，石油就会流出地表。古人很可能在有书面记录之前就发现了这一情况，并使用了石油（沥青或柏油）。我们从一些中外的古籍中知道，沥青可作为建筑物中的灰浆，可以用来填充船的缝隙，而古埃及人用其润滑战车的车轴。在美国，印第安人早在白人到来之前就把石油作为燃料。早期的美国移民不喜欢石油，因为它妨碍了从井里获得盐水。事实上，庞大的石油工业就是从瓶装药用油开始的，而那时用于照明和润滑的鲸鱼油变得越来越少。

Passage 7　Artificial Life

AL (Artifical Life) is a novel scientific pursuit that aims at studying manmade systems exhibiting behaviors that are characteristic of natural living systems. AL complements the traditional biological sciences concerned with the analysis of living organisms by attempting to synthesize life-like behaviors within computers or other artificial media. One particularly active area of artificial life is concerned with the conception and construction of artificial animals simulated by computers or by actual robots whose rules of behavior are inspired by those of animals.

Research in the field of standard artificial intelligence aims at simulating the most elaborate faculties of the human brain such as problem solving, natural language understanding, and logical reasoning.

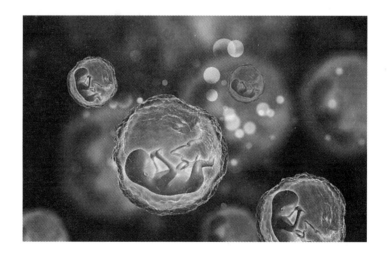

Word Bank

1. artificial [ˌɑːtɪˈfɪʃl] *adj.* 人造的；人工的
2. pursuit [pəˈsjuːt] *n.* 追赶，追求；工作　in pursuit of... 追求；追赶
3. be characteristic of sth./sb. 典型的；独特的；特有的
4. complement [ˈkɑmplɪmənt] *v.* 补充；补足

5. attempt [ə'tempt] *v.* 尝试；试图 *n.* 尝试；进攻

6. synthesize ['sɪnθəsaɪz] *v.* 合成；综合

7. elaborate [ɪ'læbərət] *adj.* 复杂的；详尽的；精心制作的 *v.* 详细说明；详尽阐述
   elaborate on 详细说明

8. conception [kən'sepʃn] *n.* 概念；构想

Exercises

I. Fill in the blanks with the words given below. Change the form where necessary.

characteristic	artificial	inherit	complement
capable	elaborate	synthesize	pursuit

1. The prisoner _____ to escape, but failed.

2. He has decorated an _____ room.

3. We _____ one another perfectly.

4. After a _____ lasting all day we finally caught up with them.

5. Raining is considered to be _____ of London.

6. The plan was brilliant in its _____ but failed because of lack of money.

II. Comprehension of the passage.

Choose the best answer to each of the following questions.

1. What is the AL according to the passage? _____

   A. A novel scientific pursuit.

   B. Manmade systems.

   C. Natural living systems.

   D. All of the above options.

2. This passage may be selected from _____.

   A. periodicals or magazines

   B. news website

   C. published articles

   D. the textbook

3. Which of the following is true according to this passage?_____

   A. AL complements the traditional chemical sciences.

   B. Author attempting to synthesize life.

   C. Human faculties might be inherited from the simplest adaptive abilities of animals.

   D. The animal approach is based on the understanding or construction of simulated animals.

4. Research in the field of standard artificial intelligence (AI) aims at _____.

   A. simulating the most elaborate faculties of the human brain

   B. problem solving

   C. natural language understanding

   D. logical reasoning

Reference translation

人工生命

人工生命是一种新兴的研究方向，旨在研究展示典型的自然生物体系行为的人造系统。人工生命补充了传统的生物科学，它通过在计算机或其他人工介质中合成类似生命的行为来分析活的生物体。人工生命有一个特别活跃的领域，涉及由计算机模拟的或行为规则受到动物启发的机器人或人工动物的概念和构造。

标准人工智能领域的研究旨在模拟人类大脑最复杂的能力，如解决问题的能力、自然语言能力理解和逻辑推理的能力。

Passage 8　3D Food Printer

Scientists have developed a commercially viable 3D food printer, which uses raw food "inks" that are fed into the printer and once you load the recipe and press the button, voila! An electronic blueprint states exactly what materials go where and are drawn up using traditional engineering CAD software. So anything that can be loaded into syringes–liquid cheese, chocolate and cake batter–can be printed out! So far, they have had some successes creating cookies, cake, sushi and "designer domes" made of turkey meat. You can imagine a 3D printer making homemade apple pie without the need for all of the materials in these processes like cars, trucks, pans, coolers, etc. 3D printing will do for food what e-mail and instant messaging did for communication.

Word bank

1. commercially [kə'mɜːʃəli] *adv.* 商业上；通商上
2. viable ['vaɪəbl] *adj.* 切实可行的；能存活的
3. CAD=computer-aided design

4. raw [rɔ:] *adj.* 生的；未加工的

5. recipe ['resəpi] *n.* 食谱；处方

6. voila ['vɔɪlə] *inj.* <法>那就是；瞧（表示事情成功或满意之感叹词用语）

7. blueprint ['blu:prɪnt] *n.* 蓝图，设计图；计划，大纲

8. syringe [sɪ'rɪndʒ] *n.* 注射器，注射筒；灌肠器
   syringes-liquid cheese 灌注奶酪，灌液奶酪

9. apple pie 苹果派，苹果馅饼

Exercises

I. Fill in the blanks with the words given below. Change the form where necessary.

| blueprint | recipe | inherit | fabricate |
| fertilize | viable | commercially | apple pie |

1. He helps his father _____ the crops.

2. I had a hamburger, some French fries, _____ and a Coke.

3. These fish are often eaten _____.

4. The new products are more difficult and expensive to _____.

5. By the example, that method is _____.

6. They are to _____ 100,000 US dollars.

II. Comprehension of the passage.

Choose the best answer to each of the following questions.

1. The theme of the essay is about_____.
   A. nature　　　　　　B. politics
   C. literature　　　　　D. science

2. This article is to explain the method of _____.
   A. direct description
   B. making assumption
   C. examplification
   D. column number

3. What is the raw material used in the 3D food printer? _____
   A. Ink.

B. Food raw materials.

C. Printing paper.

D. Various color dyes.

4. The main feature of 3D food printer is _____.

   A. it can easily produce food

   B. it eliminates the processes like cars, trucks, pans, coolers, etc.

   C. it can produce canned food

   D. it can be drawn using CAD software

Reference translation

3D 食物打印机

科学家们研发出了一款可商用的 3D 食物打印机。它以食物原料作"墨水",用户只要把食物的材料和配料预先放入容器内,再输入食谱,按下按钮,一切搞定。利用传统工程计算机 CAD 软件画出的设计图确定了原料的实际走向。因此,任何可以被装入灌注器的食物——比如灌液奶酪、巧克力和奶油煎饼——都可以打印出来。迄今为止,科学家已成功"打印"出了饼干、蛋糕、寿司以及火鸡肉制作的"特制肉塔"。你可以设想一下,使用 3D 食物打印机自制苹果派,而无须使用生产过程中用到的汽车、卡车、平底锅、冷藏器等工具。3D 食物打印机可以如即时快捷交流的电子邮件那样方便地制作出食物。

Passage 9　Next Innovations

Some innovations, including air-powered batteries, 3D cell phones as well as "adaptive traffic systems", could be expected. Today's batteries could be replaced by batteries "that use the air we breathe" and that will "last about 10 times longer than they do today". In some cases, batteries may even disappear in smaller devices.

Holographic cameras will become widespread. And 3D and holographic cameras that fit into cell phones allowing video chat with "3D holograms of your friends in real time".

"Adaptive traffic system" adopts new mathematical models and predictive analytics technologies to deliver the best routes for daily travel. It will learn traveler behavior to provide more dynamic travel safety and route information to travelers than is available today.

In addition, better ways to recycle heat and energy from data centers will be found to "heat buildings in the winter and power air conditioning in the summer" in the future.

The last but not the least, "citizen scientists" will emerge with sensors in cell phones, cars and wallets collecting data for research.

Word bank

1. innovation [ˌɪnə'veɪʃn] *n.* 创新；新观念；新发明
2. adaptive [ə'dæptɪv] *adj.* 适应的；有适应能力的
3. holographic [ˌhɒlə'græfɪk] *adj.* 全息的；全部手写的
   holographic camera 全息照相机
4. analytics [ˌænə'lɪtɪks] *n.* 分析学；解析学；分析论
5. deliver [dɪ'lɪvə(r)] *v.* 递送；发表；使分娩
6. available [ə'veɪləbl] *adj.* 可获得的；有空的；可购得的
7. recycle [ˌriː'saɪkl] *v.* 回收利用；使再循环
8. emerge [i'mɜːdʒ] *v.* 出现，浮现

Exercises

I. Fill in the blanks with the words given below. Change the form where necessary.

analytics	recycle	emerge	innovation
adaptive	available	holographic	deliver

1. There are three small boats _____ for hire.
2. I have orders to _____ it to Mrs. Doris personally.
3. This technical _____ will save us much time and labor.
4. David was waiting outside the door as she _____.
5. We _____ all our empty food and drink cans.
6. Reducing water loss is a(an) _____ feature of plant.

II. Comprehension of the passage.

Choose the best answer to each of the following questions.

1. The word "alir-powered" (Line 1) means that _____.
    A. powered by air
    B. wireless charger
    C. up in the air
    D. ipad pro
2. Which of the following is true? _____
    A. Batteries may even disappear in smaller devices.
    B. "Adaptive traffic system" adopts new mathematical models and predictive analytics and technologies.
    C. The Air-powered batteries has been used now.
    D. The Air-powered batteries is useless to our life.
3. Which characteristic is not mentioned about the future? _____
    A. Adaptive traffic system.
    B. 3D cell phones.
    C. Holographic cameras.
    D. Environment.

4. What is the author's attitude? _____
   A. Positive.
   B. Neutral.
   C. Negative.
   D. Casual.

Reference translation

未来的创新

预计未来可能会有空气动力电池、3D 手机以及"适应性交通系统"等科技创新。如今的电池将被以空气为动力的电池所替代，新型电池寿命也将是现在电池寿命的 10 倍。在某些情况下，较小的设备甚至不再需要电池。

全息摄影将得到普及。装载有 3D 全息摄像头的手机将使人们可以与好友进行实时 3D 全息视频聊天。

"适应性交通系统"使用创新的数字模型和预测分析技术为人们每天出行提供最便捷的路线。它将会了解出行者的行踪，从而提供出更动态、更安全的防护措施以及路况信息。

此外，在未来的数据中心运作时散发的热能将会被更好地循环利用，并且可以做到冬天为建筑物供暖，夏天为空调供电。

最后一点也不容小觑，"平民科学家"将在今后出现，手机、汽车以及钱包中的传感器可以随时收集数据以供研究。

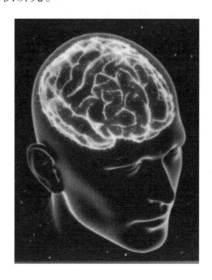

Passage 10　Globe-spanning Telescope

This super-observatory probes nearby galaxies for black holes. While no telescope can "see" a black hole by definition, the array recently provided the most <u>compelling</u> evidence yet for the existence of the phenomenon; it spotted a group of water masers, the radio equivalent of lasers, whirling around a mysterious invisible object of staggering gravitational strength in a galaxy 21 million light-years away. To find something like this so early on in the life of the array is a great and big surprise.

The array is now scrutinizing a small, phenomenally powerful object just 11,000 light-years away, a potential black hole in our galactic backyard.

Word bank

1. observatory [əb'zɜːvətri] *n.* 天文台；气象台；瞭望台
2. probe [prəʊb] *v.* 探测；探查
3. array [ə'reɪ] *n.* 数列；阵列；大量
4. compelling [kəm'pelɪŋ] *adj.* 令人信服的；引人入胜的；扣人心弦的
5. maser ['meɪzə] *n.* 磁波；微波激射器
6. whirl [wɜːl] *v.* 旋转；回旋

7. staggering ['stægərɪŋ] *adj.* 蹒跚的；难以置信的
8. galaxy ['gæləksi] *n.* 星系；银河系；*pl.* galaxies
9. scrutinize ['skru:tənaɪz] *v.* 仔细查看；仔细检查
10. phenomenally [fə'nɒmɪnəli] *adv.* 极其；非常；非凡的；现象上地；明白地

Exercises

I. Fill in the blanks with the words given below. Change the form where necessary.

| scrutinize | probe | phenomenally | compelling |
| whirl | galaxy | capture | staggering |

1. Leaves _____ in the wind.
2. There is no _____ reason to believe him.
3. Astronomers have discovered a distant _____.
4. The doctor _____ the wound for signs of infections.
5. I saw a drunk _____ up the street.
6. The statement was carefully _____ before publication.

II. Comprehension of the passage.

Choose the best answer to each of the following questions.

1. Which of the following has the closest meaning with the underline word "compelling"? _____
   A. Compressing.
   B. Comprehensive.
   C. Convincing.
   D. Magnificent.

2. The word "this" in the sentence "to find something like this so early..." refers to _____.
   A. an object that the array sees
   B. a galaxy 21 million light-years away
   C. a group of water masers whirling around a mysterious invisible object
   D. the gravitational strength of a galaxy

3. What's the attitude of Phillip Diamond toward the new finding? _____
   A. Positive.
   B. Negative.
   C. Neutral.
   D. Pessimistic.

4. It can be inferred from this passage that _____.
   A. black holes only exist in a galaxy 21 million light-years away
   B. only the super-observatory can see black holes
   C. nobody was surprised when the radio telescope found the phenomenon
   D. no instrument ever before has found more convincing evidence of the existence of black holes than the super-observatory

Reference translation

太空望远镜

这个"超级望远镜"探测附近的星系，寻找黑洞。虽然没有望远镜可以"探测"到所谓的黑洞，但该阵列最近提供了迄今为止最有力的证据，证明了这一现象的存在。它发现了一组水微波激射器，这是一种无线电当量的激光，围绕着一个在2 100万光年之外的星系中一个神秘的隐形物体旋转。在阵列生命的早期发现这样的东西是一个巨大的惊喜。

现在，该阵列正在检查距离我们11 000光年远的一个非常强大的小型物体，这是我们银河系的一个潜在的黑洞。

Passage 11　Bionic Leaf

The scientists craft a kind of living battery and call it bionic leaf for its melding of biology and technology. The device uses solar electricity from a photovoltaic panel to power the chemistry that splits water into oxygen and hydrogen. Microbes within the system then feed on the hydrogen and convert carbon dioxide in the air into alcohol that can be burned as fuel.

The researchers found the new catalyst in an alloy of cobalt and phosphorus. With this new catalyst in the bionic leaf, the team boosted efficiency at producing alcohol fuels such as isopropanol and isobutanol to roughly 10 percent. In other words, for 1kw of electricity, the microbes could scrub 130g of CO_2 out of the air to make 60g of isopropanol fuel. Such a conversion is roughly 10 times more efficient than natural photosynthesis.

Word bank

1. bionic [baɪ'ɒnɪk] *adj.* 仿生学的
2. craft [krɑːft] *v.* 精心制作；周密制订 *n.* 手艺；工艺
3. meld [meld] *v.* （使）融合，合并
4. photovoltaic [fəʊtəʊvɒl'teɪɪk] *adj.* 光电池的
5. split [splɪt] *v.* 分裂；分开
6. catalyst ['kæt(ə)lɪst] *n.* 催化剂；刺激因素
7. boost [buːst] *v.* 促进，提高
8. scrub [skrʌb] *v.* 使净化；用力擦洗　scrub...off...（用刷子等）刷掉，擦掉
   scrub...out...（用刷子等）把某物从里到外擦洗干净
9. conversion [kən'vɜːʃn] *n.* 变换，转变
10. photosynthesis [ˌfəʊtəʊ'sɪnθəsɪs] *n.* 光合作用

11. isopropanol [ˌaɪsə'prəʊpənəʊl] *n.* 异丙醇

12. isobutanol [ˌaɪsɒb'juːtaːnɒl] *n.* 异丁醇

Exercises

I. Fill in the blanks with the words given below. Change the form where necessary.

| scrub | catalyst | split | bionic |
| meld | boost | conversion | craft |

1. We need a holiday to _____ our spirits.
2. Her company has had to _____ up and work from two locations.
3. The corridors are _____ clean.
4. A _____ is a substance which speeds up a chemical reaction.
5. He plans to _____ the other two steel plants into his.
6. The windows would have been _____ in the latter part of the 18th century.

II. Comprehension of the passage.

Choose the best answer to each of the following questions.

1. The bionic leaf is _____.
   A. a common leaf
   B. a machine which can make oxygen
   C. a living battery for its melding of biology and technology
   D. microbes feed on the hydrogen

2. Which of the following is not true? _____
   A. The bionic leaf needs solar electricity.
   B. The principle of the bionic leaf is the same as the nature leaf.
   C. The bionic leaf has many microbes.
   D. The bionic leaf will benefit environment.

3. What is the meaning of the "catalyst" in the passage? _____
   A. In chemistry, a catalyst is a substance that causes a chemical reaction to take place more quickly.
   B. A person that causes a change.
   C. A person that causes event to happen as a catalyst.

D. It boosted efficiency at producing alcohol fuels.

4. What do we learn from the passage? _____

    A. Microbes of the bionic leaf feed on the oxygen.

    B. Microbes can scrub CO_2 of the air and isopropanol fuel.

    C. The bionic leaf could help mitigate planet-warming pollution problems.

    D. The bionic leaf is widely used by people.

Reference translation

仿生叶

    科学家精心研制了一种生物电池，并且称它是生物和科技相结合的仿生叶。这个装置利用太阳能光伏板发电，将水分解为氧气和氢气。系统内微生物以氢气为原料，将空气中二氧化碳转化为可以作为燃料燃烧的酒精。

    研究人员在钴和磷的合金中发现了新的催化剂。仿生叶有了这个新的催化剂，研究团队提高了生产酒精燃料的效率，如异丙醇、异丁醇提高了10%左右。换句话说，微生物用1千瓦的电力可以从空气中除去130克的二氧化碳并产生异丙醇燃料60克。这种转化效率大约是自然光合作用的10倍。

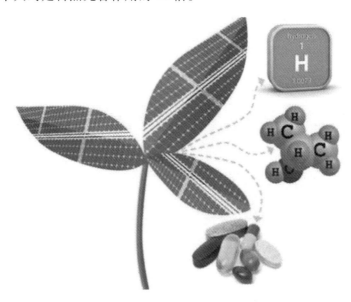

Passage 12　Brain Damage

A new study revealed cognitive impairment in mice when subjected to highly ionized radiation. This radiation is similar to what astronauts will face on deep space missions, like one to Mars.

Astronauts on a mission to Mars will be expected to rely on their memory, multitasking and decision-making skills to handle the first manned expedition beyond the moon. But the harmful radiation from galactic cosmic rays throughout their journey could cause them to experience cognitive impairment in these key areas, as well as increased depression and anxiety, according to a new study.

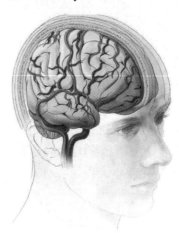

The effects of exposure to highly charged and ionized particles during extended deep-space travel could be long-lasting and without resolve, similar to dementia.

Word bank

1. cognitive ['kɒgnətɪv] *adj.* 认知的；认识的
2. impairment [ɪm'peəmənt] *n.* 损害；损伤
3. subjected to　使遭受
4. ionize ['aɪənaɪz] *v.* （使）电离；（使）成离子
5. radiation [ˌreɪdi'eɪʃn] *n.* 辐射；放射物
6. mission ['mɪʃn] *n.* 使命；任务；代表团
7. galactic [gə'læktɪk] *adj.* 银河的　galaxy *n.* 银河系；星系
8. cosmic ['kɒzmɪk] *adj.* 宇宙的；极广阔的
9. particle ['pɑːtɪkl] *n.* 微粒，颗粒

Exercises

I. Fill in the blanks with the words given below. Change the form where necessary.

mission	radiation	ionize	galactic
dementia	cognitive	impairment	cosmic

1. A _____ of something is a very small piece or amount of it.
2. There is an enormous sense of _____ in his speech and gesture.
3. He has a visual _____ in the right eye.
4. Do you believe in a _____ plan?
5. She is suffering from _____.
6. As children grow older, their _____ processes become sharper.

II. Comprehension of the passage.

Choose the best answer to each of the following questions.

1. Astronauts have cognitive impairment because of _____.
    A. weightlessness
    B. harmful effects of radiation
    C. unhealthy body
    D. fear of space
2. The astronauts on Mars feel like _____.
    A. just born baby
    B. students in a doze
    C. the old with dementia
    D. suffering from depression
3. The harmful effects of radiation could cause astronauts to experience _____.
    A. occasional memory loss
    B. sleeplessness
    C. cognitive impairment
    D. suffering from dementia immediately
4. Which of the following statement is true? _____
    A. On Mars, the astronauts can rely on their memory, multitasking and decision-making capabilities to deal with the expedition.

B. Astronauts on Mars are similar to patients with dementia.
C. Man has landed on Mars.
D. Highly ionizing radiation only affects rats and people.

Reference translation

脑部损伤

一项新的研究表明,受到高度电离辐射的老鼠会出现认知障碍。这种辐射类似于宇航员在深太空,比如在火星执行任务时受到的辐射。

科学家期望去火星执行任务的宇航员们能够依靠他们的记忆、多任务处理能力和决策能力来完成除登月以外的第一次载人探险。但是一项新的研究显示,在整个旅程中,来自银河系宇宙射线的有害辐射可能会导致他们在这些关键领域的认知障碍,以及引发更多抑郁和焦虑的情绪。

在长时间的深太空旅行中,暴露在高电荷和离子化的粒子下的影响,就像痴呆症一样,效果可能是持久的并且不会消退。

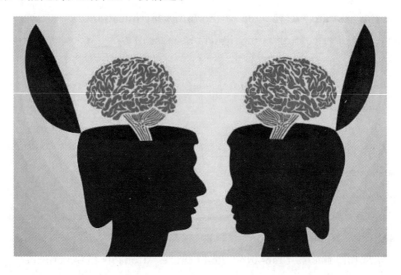

Passage 13　Changing Smart Phone into a Microscope

The researchers discovered a new kind of optical lens recently which can be stuck directly on the smart phone. This optical lens which resolution is up to 1 micron can magnify the image 100 times. However, it costs only 3 cents. It's said that this lens is made of a material named PDMS which concentration is similar to honey. The PDMS can also accurately be attached on a preheating surface and gradually solidified. What's more, the PDMS and glass will not permanently stick together. As a result, we can easily get the lens off after using it.

Word bank

1. optical ['ɒptɪkl] *adj.* 光学的；视觉的
2. lens [lenz] *n.* (*pl.* lenses) 透镜；镜头
3. stick on 贴上；保持在……之上
4. resolution [rezə'lu:ʃ(ə)n] *n.* 分辨率；解决；决心
5. up to 多达；由……决定
6. micron ['maɪkrɒn] *n.* 微米

7. magnify ['mægnɪfaɪ] *v.* 放大；夸大

8. concentration [ˌkɒnsn'treɪʃn] *n.* 浓度；专心，专注

9. be similar to 与……相似

10. accurately ['ækjərətlɪ] *adv.* 精确地；准确地；正确无误地

11. permanently ['pɜːmənəntlɪ] *adv.* 永久地；长期不变地

Exercises

I. Fill in the blanks with the words given below. Change the form where necessary.

| magnify | permanently | optical | stick on |
| concentration | lens | accurately | up to |

1. What is the _____ of salt in sea water?
2. He has settled _____ in the states.
3. Telescopes and microscopes are _____ instruments.
4. The _____ of a camera forms images.
5. What can I _____ this _____ with?
6. I can _____ express myself in English.

II. Comprehension of the passage.

Choose the best answer to each of the following questions.

1. Which place do you think needn't to use this kind of lens? _____
   A. Hospital.
   B. School.
   C. Factory.
   D. Family.

2. Which of the following do you think is not right? _____
   A. PDMS is a new kind of material.
   B. This optical lens is very cheap.
   C. This optical lens is soft.
   D. The smart phone and microscope can be separated.

3. Which one can be inferred from the article? _____
   A. The PDMS cannot change its state.

B. We can change a big phone into a microscope like a magic.

C. Primary school students can afford the lens to do some experiments.

D. We can buy the lens right now.

4. Which of the following sentence describes the material of PDMS? _____

A. The optical lens can be stuck directly on the smart phone.

B. This optical lens can magnify the image 100 times.

C. The PDMS can accurately be attached on a preheating surface and gradually solidified.

D. It costs only 3 cents.

Reference translation

智能手机变成显微镜

研究人员近日开发出一种新型光学镜头,能直接贴在智能手机上,将图像放大100倍,分辨率达到1微米,而成本只需要3美分。据报道,这种镜头用二甲硅氧烷(PDMS)制成。PDMS是一种浓度像蜂蜜的材料,能精确附着在一个预热表面逐渐凝固。而且,由于PDMS和玻璃不会永久地粘在一起,所以用过镜头后还可以轻易取下来。

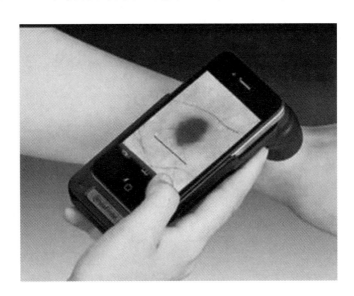

Passage 14　Disabled People Can Move Again

Earlier this year, it was reported using a brain machine interface to allow a quadriplegic to move his finger and arms. That system directly activated arm muscles with brain signals detected by implanted electrodes, bypassing the damaged spinal cord. The training strategy is less invasive but would likely be hard to fully extend to a large number of paralyzed people given the expense of the custom exoskeleton. The first two stages—the virtual reality and treadmill systems—could be more easily adopted, although it's unclear how much recovery they promoted. The researcher suggests that brain-machine interfaces could one day be combined with several other treatment strategies—such as stem cells and drugs—to produce much greater overall benefit for people with damaged spinal cords. However no single strategy will be enough to overcome severe paralysis. So it's appropriate to be cautious.

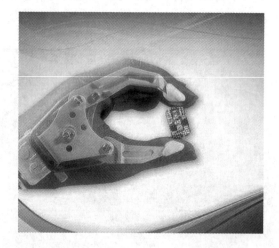

Word bank

1. interface ['ɪntəfeɪs] *n.* 界面；[计]接口；交界面
2. quadriplegic [ˌkwɒdrɪ'pliːdʒɪk] *n.* 四肢瘫痪者　*adj.* 四肢瘫痪的
3. activate ['æktɪveɪt] *v.* 使活动；使开始作用
4. implant [ɪm'plɑːnt] *v.* 移植；植入；灌输

5. bypass ['baɪpɑːs] *v.* 疏通；忽视 *n.* 旁道支路

6. invasive [ɪn'veɪsɪv] *v.* 侵略性的；侵害的 less invasive 微创

7. paralyze ['pærəlaɪz] *v.* 使瘫痪；使麻痹；使不能正常活动

8. exoskeleton ['eksəuskelɪtn] *n.* 外骨骼

9. treadmill ['tredmɪl] *n.* 踏车；跑步机

10. appropriate [ə'prəuprɪət] *adj.* 恰当的；合适的

Exercises

I. Fill in the blanks with the words given below. Change the form where necessary.

| quadriplegic | implant | activate | appropriate |
| interface | paralyze | bypass | pursuit |

1. Jeans are not _____ for a formal party.
2. The old man had a heart _____ surgery two weeks ago.
3. He is now _____ and confined to a wheelchair.
4. You might _____ it by a button.
5. The operation to _____ the artificial heart took two hours.
6. The new version of the program comes with a much better user _____ than the original.

II. Comprehension of the passage.

Choose the best answer to each of the following questions.

1. This system includes _____.
   A. a pair of new arm muscles
   B. a pair of new spinal cords
   C. implanted electrodes connected with brain
   D. stem cells and drugs

2. Now this system can help disabled people _____.
   A. move part of their disabled bodies
   B. do sports like normal people
   C. be stronger than normal people
   D. make their bodies immortal

3. In the future, disabled people can get fully recovered _____.
   A. just by this system
   B. with this system and other treatment strategies
   C. with stem cells and drugs
   D. with custom exoskeleton

4. From the report we know that _____.
   A. with the help of this system, disabled people can move like the others now
   B. this technology still needs further development
   C. disabled people can be cured by drugs
   D. disabled people can't move the same as others anyhow

Reference translation

残疾人可恢复行动了

据今年早些时候的报道，使用和大脑连接的计算机成功地帮助了一个四肢瘫痪者移动他的手指和手臂。该系统通过植入电极绕过受损脊髓，直接激活手臂肌肉与脑信号。尽管这种康复疗法是微创，但考虑到定制外骨骼的费用，可能很难大量推广到瘫痪的患者。头两个阶段——虚拟现实和跑步系统——可能更容易被采用，尽管它们促进恢复的程度目前还不清楚。研究者建议，脑机互动系统有朝一日可以与其他几种治疗方法（如干细胞和药物）相结合，为脊髓损伤患者带来更大的好处。但是没有一种单一的方法能够克服严重的瘫痪，所以谨慎一些是应该的。

Passage 15　First Quantum Satellite

China successfully launched the world's first quantum satellite from the Jiuquan Satellite Launch Center in northwest China's Gansu Province on August 16, 2016. The 640-kg satellite, known as Quantum Experiments at Space Scale (QUEES), is nicknamed Micius after a fifth-century B.C. Chinese philosopher and scientist credited as the first person ever to conduct optical experiments.

Having entered a sun-synchronous orbits at an altitude of 500 km, the satellite circles Earth once every 90 minutes. During its two-year mission, QUESS is designed to establish "hack-proof" quantum communication by transmitting uncrackable keys and to provide insights into quantum entanglement, the strangest phenomenon in quantum physics. Quantum communication provides ultra-high security. As quantum photons can neither be separated nor duplicated, it is impossible to intercept or decipher information thus encoded and transmitted.

Word bank

1. quantum ['kwɒntəm] *n.* (*pl.*) quanta [物]量子；定量；总量
2. nickname ['nɪkneɪm] *v.* 给……起绰号；*n.* 昵称；绰号
3. credit ['kredɪt] *v.* 相信，信任；归功于 be credited as 被誉为

4. subjected to 使遭受
5. sun-synchronous orbits 太阳同步轨道
6. crackable [k'rækəbl] *adj.* 会裂开的，会粉碎的 uncrackable 无法破解的
7. entanglement [ɪn'tæŋglmənt] *n.* 瓜葛；牵连；纠缠
8. ultra-high security 极高的安全性
9. duplicate ['djuːplɪkeɪt] *n.* 复制；复印；重复
10. intercept [ˌɪntə'sept] *v.* 拦截，拦住
11. decipher [dɪ'saɪfə(r)] *v.* 破译（密码）；辨认（潦草字迹）
12. encode [ɪn'kəʊd] *v.* （将文字材料）译成密码；编码

Exercises

I. Fill in the blanks with the words given below. Change the form where necessary.

| intercept | duplicate | encode | entanglement |
| nickname | credited | decipher | quantum |

1. I'm still no closer to _____ the code.
2. Red got his _____ for his red hair.
3. It is illegal to _____ radio messages.
4. The book is _____ as one of the world's most influential works.
5. Scientists hope the work may be _____ elsewhere.
6. Many dolphins die each year from _____ in fishing nets.

II. Comprehension of the passage.

Choose the best answer to each of the following questions.

1. The purpose of QUESS is to _____.
   A. accelerate the speed of satellite
   B. add up the safety of quantum communication
   C. eliminate the shortcoming of quantum communication
   D. protect the circles of satellite

2. What does the word "hack-proof" (Line 3, Para. 2) possibly mean? _____
   A. The protection of one's body.
   B. A new roof of one's house.

C. The defense of one's computer or phone.

D. The prove of some questions.

3. The author's attitude towards Micius is _____.

   A. favorable

   B. neutral

   C. doubtful

   D. critical

4. Which of the following statements is TRUE? _____

   A. Quantum photons can be copied.

   B. Quantum photons can transport information.

   C. Quantum photons can head off information inevitably.

   D. Quantum photons can't gain the right information.

Reference translation

第一颗量子卫星

中国于 2016 年 8 月 16 日在中国西北甘肃省酒泉卫星发射中心成功发射了世界上第一颗量子卫星。这颗因空间尺度量子实验而著名的重达 640 千克的卫星以"墨子"的名字命名——墨子被认为是公元前 5 世纪中国第一个开创了光学实验的哲学家和科学家。

卫星已经在 500 千米高空进入了太阳同步轨道，其绕地球运转一周需时 90 分钟。在量子卫星两年的任务期间，空间尺度量子实验设计的初衷是通过发射不可破解的密匙来建立"防黑客"量子通讯，并提供对量子纠缠——量子物理中最奇特现象的见解。量子通讯提供超高的安全性。因为量子光既不能被分割也不能被复制，因此它不可能被拦截或破译信息从而编码及传送它们。

Passage 16 Lie Detector App

A technology company is developing a lie detector app for smart phones that could be used by parents, teachers and internet daters.

The app measures blood flow in the face to assess whether or not you are telling the truth. Its developers say that it can be used for daters wanting to see if somebody really is interested in them. The idea is that different human emotions create different facial blood flow patterns. These patterns change if we are telling the truth or telling a lie. Using footage from the smart phone camera, the software will see the changes in skin color and compare them to standardized results. A study from last year found that anger was associated with more blood flow and redness whilst sadness was associated with less of both. Developmental neuroscientist said that the lie detector test will let you find out the truth non-invasively, and sometimes it can be covertly.

Word bank

1. detector [dɪ'tektə(r)] *n.* 探测器；发现者 a smoke detector 烟雾检测仪
2. flow [fləʊ] *n.* 流动 blood flow 血流
3. footage ['fʊtɪdʒ] *n.* （影片中的）连续镜头；片段
4. standardized ['stændədaɪzd] *adj.* 标准的；定型的
5. associated with 联系；与……交往
6. whilst [waɪlst] *conj.* 同时；当……的时候；而
7. neuroscientist ['njʊərəʊsaɪəntɪst] *n.* [医]神经科学家；神经科学各分支的专家
8. invasively [ɪn'veɪsɪvlɪ] *adv.* 侵入式 non-invasively 非侵入式
   invasive cancer 扩散性肿瘤
9. covertly ['kʌvətlɪ] *adv.* 偷偷地；暗中地；秘密地

Exercises

I. Fill in the blanks with the words given below. Change the form where necessary.

| non-invasively | footage | covertly | neuroscientist |
| capable | whilst | detector | flow |

1. I looked out to the kitchen _____.
2. The parts of an automobile are _____.
3. The first two services are free, _____ the third costs £35.00.
4. She tried to stop the _____ of blood from the wound.
5. Blood pressures can be measured invasively or _____.
6. For decades, no _____ has been known to repeat the experiment.

II. Comprehension of the passage.

Choose the best answer to each of the following questions.

1. Lie detector app can test whether or not people telling the truth by _____.
   A. people's voice
   B. people's action
   C. people's blood flow on the face
   D. people's expression
2. Which of the following is correct according to the article? _____
   A. The lie detector app is now promoted.
   B. The lie detector app needs the camera to run.
   C. The lie detector app can only be used for entertainment.
   D. The lie detector app's creation idea came from illusion.
3. The author's view of this app is _____.
   A. just by this system
   B. exaggerated
   C. objective
   D. favorite
4. Which of the following statements can prove that the principle of the lie detector has certain scientific basis?
   A. Because different human emotions create different facial blood flow patterns, these patterns change if we are telling the truth or telling a lie.

B. The software will see the changes in skin color and compare them to standardized results.
C. The lie detector app for smart phones could be used by parents, teachers and internet daters.
D. Anger was associated with more blood flow and redness whilst sadness was associated with less of both.

Reference translation

谎言探测应用程序

一家技术公司正在开发智能手机的测谎应用程序，可供家长、教师和约会的网友使用。

该应用程序通过测量脸上的血流量来评估你是否说谎。开发人员说，用这个应用程序可以知道约会对象对自己是否真的感兴趣。其原理是人们不同的感情产生不同的面部血液流动模式。这些模式会因我们说的是真话或谎话而改变。使用智能手机的摄像头，该软件将看到皮肤颜色的变化，并将其与标准结果进行比较。去年的一项研究发现，愤怒常伴随血流量的增加和面部发红，而悲伤则与此相反。发育神经科学家指出，测谎仪能够以一种"安全远程，时而隐秘"的方式发现真相。

Passage 17　A New Skin Sensor

If you ever searched a way to quantity perspiration, this new tech might interest you. Researchers have created a soft adhesive path that can measure the composition of sweat. You can scan it with smartphone and an App will give information about electrolyte balance, dehydration levels and total water loss. It works like this: As your pores release sweat, a ring-shaped channel fills up and diverts into four different sensors that absorb the moisture. Each sensor has a corresponding color—blue, yellow, orange or red, and each one measures something different: chloride, glucose, pH or lactate. The color becomes more vibrant based on the concentration of what it monitors. Measuring electrolyte loss can combat fatigue, and tracking chloride ions can indicate susceptibility for diseases. With a little tweaking, it could even be used to test for doping at athletic events.

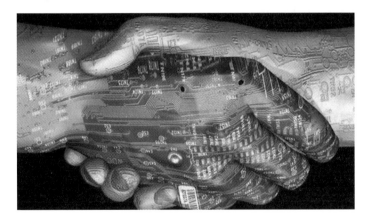

Word bank

1. perspiration [ˌpɜːspəˈreɪʃn] *adj.* 汗；汗珠；出汗
2. adhesive [ədˈhiːsɪv] *n.* 黏合的；有附着力的
3. electrolyte [ɪˈlektrəlaɪt]　*n.* [化]电解液，电解质
4. dehydration [ˌdiːhaɪˈdreɪʃn] *n.* 脱水；失水
5. pore [pɔː(r)] *n.* 毛孔；细孔
6. divert [daɪˈvɜːt] *v.* 使转移；使分心
7. chloride [ˈklɔːraɪd] *n.* 氯化物

8. glucose ['glu:kəʊs] *n.* 葡萄糖

9. lactate [læk'teɪt] *n.* 乳酸盐

10. track [træk] *v.* 监测；跟踪 *n.* 小路；路线

11. susceptibility [sə,septə'bɪləti] *n.* 易受影响或损害的特性；敏感性

12. tweak [twi:k] *v.* 拧；扭；稍稍调整（机器、系统等）

Exercises

I. Fill in the blanks with the words given below. Change the form where necessary.

| pores | dehydration | tweak | lactate |
| divert | track | synthesize | perspiration |

1. From this interface you can manage, _____ the server.
2. A plant's lungs are the microscopic _____ in its leaves.
3. Lack of water can lead to _____.
4. His hands were wet with _____.
5. Don't _____ the subject into a side issue.
6. Would you give me an _____ stamp?

II. Comprehension of the passage.

Choose the best answer to each of the following questions.

1. A skin sensor is _____.
   A. a soft adhesive path
   B. a medical equipment
   C. a monitor
   D. an app

2. According to the passage, a skin sensor can _____.
   A. give you information about your health
   B. test your body's doping
   C. measure the composition of you sweat
   D. combat fatigue, indicate susceptibility and release perspiration

3. Which of the following statements is true? _____
   A. The skin sensor can fill up your ring-shaped channel, and diverts four different

sensors.

B. The color of the skin sensor becomes different based on the concentration of what it monitors.

C. The skin sensor can test for doping now.

D. The skin sensor can measure electrolyte and track chloride ions.

4. What does the "it" mean in the last sentence? _____

A. Path.

B. Monitor.

C. Perspiration.

D. Sweat.

Reference translation

新型皮肤传感器

如果你想寻找一种方法来量化你的汗水，那么这个新技术可能会引起你的兴趣。研究人员发明了一种可以测量汗液成分的软性黏合剂方法。你可以用智能手机扫描它，应用程序会提供电解质平衡、脱水水平和总失水的信息。它是这样工作的：当你的毛孔释放汗水，环形通道填满，将分为四个不同的传感器吸收水分。每个传感器都有一个对应的颜色——蓝色、黄色、橙色或红色，每一个都测量不同的东西：氯化物、葡萄糖、pH 值或乳酸。根据显示器指示的浓度，颜色变得更加有活力。测量电解质损耗可以抵抗疲劳，跟踪氯离子可以表明疾病的易感性。只要稍微调整一下，它甚至可以用来测试体育比赛中的兴奋剂情况。

Passage 18   Quinine

Quinine is a medication to treat malaria. It is on the WHO Model List of Essential Medicines, the most important medications needed in a basic health system. Quinine was first isolated in 1820 from the bark of a cinchona tree.

The Jesuits were the first to bring cinchona to Europe. The Spanish were aware of the medicinal properties of cinchona bark by the 1570s or earlier: Nicolás Monardes (1571) and Juan Fragoso (1572) both described a tree that was subsequently identified as the cinchona tree and whose bark was used to produce a drink to treat diarrhea. Quinine has been used by Europeans since at least the early 17th century. Quinine also played a significant role in the colonization of Africa by Europeans. A historian has stated, "it was quinine's efficacy that gave colonists fresh opportunities to swarm into the Gold Coast, Nigeria and other parts of west Africa."

Word bank

1. malaria [mə'leəriə] *n.* [医]疟疾；瘴气
2. cinchona [sɪŋ'kəʊnə] *n.* (产于南美洲的) 金鸡纳树
3. medicinal [mə'dɪsɪnl] *adj.* 医学的；医药的
4. property ['prɒpəti] *n.* 特性；属性；财产
5. bark [bɑːk] *n.* 树皮；犬吠

6. subsequently ['sʌbsɪkwəntli] *adv.* 其后；随后

7. be identified as 被识别为……，被确认为……

8. diarrhea [ˌdaɪə'rɪə] *n.* 腹泻

9. colonization [ˌkɒlənaɪ'zeɪʃn] *n.* 殖民地的开拓；殖民，殖民地化

   colonist ['kɒlənɪst] *n.* 殖民者；移民

10. efficacy ['efɪkəsi] *n.* 功效；效力

11. swarm [swɔːm] *v.* 挤满；涌往 swarm into 涌入

Exercises

I. Fill in the blanks with the words given below. Change the form where necessary.

| swarm | efficacy | bark | colonization |
| cinchona | medicinally | property | subsequently |

1. Compare the physical _____ of the two substances.
2. These countries took part in the _____ of Africa.
3. I _____ learned he wouldn't be coming.
4. Many students _____ into the canteen after class.
5. _____ may be used as raw material for paper-making.
6. Root ginger has been used _____ for centuries.

II. Comprehension of the passage.

Choose the best answer to each of the following questions.

1. This article proves that _____.

   A. quinine could treat diarrhea

   B. quinine had been used since at least the early 17th century

   C. quinine promoted the colonization of Africa

   D. there are plenty of cinchona in the Gold Coast, Nigeria and other parts of west Africa

2. When did quinine formally used as a medicine?_____

   A. In 1820.

   B. In 1570.

   C. In 1572.

D. In the early 17th century.

3. The historian's statement show that _____.

   A. European swarm into Africa for gold

   B. there was fresh air in the Gold Coast, Nigeria and other parts of west Africa

   C. the Gold Coast, Nigeria and other parts of west Africa were badly in need of quinine

   D. quinine put other parts of Africa into danger

4. We can infer that quinine _____.

   A. was the bark of cinchona

   B. played an important role in basis health system

   C. will be widely used in many fields

   D. was used to produce a drink

Reference translation

奎 宁

奎宁是一种治疗疟疾的药物。它在世界卫生组织基本药物标准清单上，是基本卫生系统所需的最重要的药物。奎宁在1820年首次从金鸡纳树的树皮中提取出来。

耶稣会会士首先把金鸡纳树带到了欧洲。16世纪70年代或更早，西班牙人就知道金鸡纳树皮的药用性质：Nicolás Monardes（1571）和 Juan Fragoso（1572）都描述了这样一棵树，后来被鉴定为金鸡纳树，树皮用于生产饮料以治疗腹泻。至少在17世纪初以来，奎宁已被欧洲人使用。奎宁在欧洲人对非洲的殖民化方面也发挥了重要作用。一位历史学家说："正是奎宁的功效，使得殖民者有机会涌入到黄金海岸、尼日利亚和西非其他地区。"

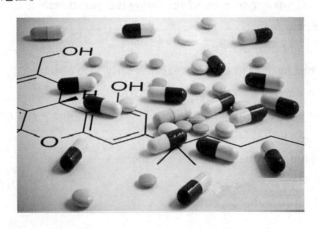

Passage 19　Renewable Biofuels from Microalgae

Rising oil prices and concerns over climate change have resulted in more emphasis on research into renewable biofuels from microalgae. Unlike plants, green microalgae have higher biomass productivity, will not compete with food and agriculture, and do not require fertile land for cultivation. However, microalgae biofuels currently suffer from high capital and operating costs due to low yields and costly extraction methods. Microalgae grown under optimal conditions produce large amounts of biomass but with low neutral lipid content, while microalgae grown in nutrient starvation accumulate high levels of neutral lipids but are slow growing. Producing lipids while maintaining high growth rates is vital for biofuel production because high biomass productivity increases yield per harvest volume while high lipid content decreases the cost of extraction per unit product. Therefore, there is a need for metabolic engineering of microalgae to constitutively produce high amounts of lipids without sacrificing growth.

Word bank

1. biofuel ['baɪəʊfjuːəl] *n.* 生物燃料
2. lipid ['lɪpɪd] *n.* [生化]脂质

3. microalgae [maɪkrəʊˈældʒiː] n. 微藻类（指肉眼看不见的藻类）

4. biomass [ˈbaɪəʊmæs] n. （单位面积或体积内）生物的数量

5. fertile [ˈfɜːtaɪl] adj. 肥沃的；可繁殖的

6. yield [jiːld] n. 产量；产额 v. 生产；屈服

7. extraction [ɪkˈstrækʃn] n. [化]提取（法）；取出；抽出

8. optimal [ˈɒptɪməl] adj. 最佳的；最适宜的；最理想的

9. nutrient [ˈnjuːtriənt] adj. 营养的；滋养的 n. 营养物

10. starvation [staːˈveɪʃn] n. [U] 饥饿；饿死；挨饿

11. accumulate [əˈkjuːmjəleɪt] v. 堆积；积累

12. metabolic [ˌmetəˈbɒlɪk] adj. 新陈代谢的；变化的

Exercises

I. Fill in the blanks with the words given below. Change the form where necessary.

nutrients	fertile	yield	biofuel
accumulated	starvation	biomass	optimal

1. I seem to have _____ a lot of books.
2. Plants draw minerals and other _____ from the soil.
3. The weight of living organisms is called _____.
4. Aim to do some physical activity three times a week for _____ health.
5. Anything grows in this _____ ground.
6. Polluted water lessens crop _____.

II. Comprehension of the passage.

Choose the best answer to each of the following questions.

1. What is the advantage of microalgae as biofuel? _____

　　A. It can product high amounts of lipids.

　　B. It is benefit to environment compared with fossil fuel.

　　C. It can product more economic benefit.

　　D. It has higher biomass productivity.

2. What is the main difference between plants and microalgae? _____

　　A. Output.

B. Growing environment.
  C. Light.
  D. Soil.
3. The author's attitude is _____.
  A. subjective   B. objective
  C. neutral   D. critical
4. Next then, what should scientists do according to this article?_____
  A. Improve biomass.
  B. Decrease the cost.
  C. Concern the climate change.
  D. Produce lipids while maintaining high growth rates.

Reference translation

可再生的微藻生物燃料

油价的上涨和对气候变化的担忧,使得人们对微藻可再生生物燃料研究更加重视。与植物不同,绿色微藻具有较高的生物量生产力,不会与粮食和农业竞争,不需要肥沃的土地种植。然而,由于低产量和高成本提取方法,微藻生物燃料目前受到投资大和运营成本高的困扰。在最佳条件下生长的微藻产生大量的生物量,但是中性脂质含量较低,而生长在低营养条件下的微藻类产生大量的中性脂质,但生长速度缓慢。在保持高生长速率的同时脂质对于生物燃料生产至关重要,因为高生物量生产率增加了单位体积的获取量,而高脂质含量降低了单位产品的提取成本。因此, 在不改变生长的条件下,有必要提高微藻类代谢工程中脂质的产量。

Passage 20 Solar Water Disinfection

Solar water disinfection is one of the most practical and low-cost techniques to reduce the load of pathogenic microorganisms in water at households in low-income areas. This method was known and used nearly 2000 years ago by Indian communities to purify drinking water. The bactericidal effect of sunlight was experimentally studied for the first time by Downes & Blunt in 1877. The first successful application was demonstrated by Acra et al. in 1980. The mechanism of action of this photo-induced process has been intensively investigated since nineties till now. Most of research was focused on the influence of key parameters like water temperature, solar energy dose, turbidity of polluted water, dissolved oxygen, nature of the microorganism, and other studies on the practical application of this technique and field trials in several locations.

Word bank

1. disinfection [ˌdɪsɪn'fekʃən] *n.* 消毒；杀菌
2. pathogenic ['pæθə'dʒenɪk] *adj.* 致病的
3. purify ['pjʊərɪfaɪ] *v.* 使纯净；净化
4. bactericidal [bækˌtɪərɪ'saɪdl] *adj.* 杀菌的

5. intensively [ɪn'tensɪvlɪ] *adj.* 强烈地；集中地
6. focus on 使聚焦于……；对（某事或做某事）予以注意
7. parameter [pə'ræmɪtə(r)] *n.* 参数；范围；限制因素
8. dose [dəʊs] *n.* 剂量；药量
9. turbidity [tɜː'bɪdətɪ] *n.* 浊度；混浊度
10. dissolve [dɪ'zɒlv] *v.* 使溶解；使液化

Exercises

I. Fill in the blanks with the words given below. Change the form where necessary.

| intensively | purify | dose | parameter |
| turbidity | focus on | dissolve | disinfection |

1. Consumers expect and have a right to demand _____-free water.
2. Mary's parents opted to educate her _____ at home.
3. Heat gently until the sugar _____.
4. Our meeting _____ the question of women's right.
5. Only _____ water is used.
6. Repeat the _____ after 12 hours if necessary.

II. Comprehension of the passage.

Choose the best answer to each of the following questions.

1. _____ was studied during the first experiment.
   A. The component of sunlight
   B. The bactericidal effect of sunlight
   C. The use of sunlight
   D. The scatter of sunlight
2. The first successful application be demonstrated _____.
   A. in 1980   B. in 2000
   C. in 1877   D. in 1880
3. Solar water disinfection aimed at _____.
   A. high-income areas   B. middle-income areas
   C. low-income areas   D. all areas

4. This passage is a piece of _____.
   A. exposition   B. argumentation
   C. narration   D. prose

Reference translation

太阳能水消毒

太阳能水消毒是减少低收入地区家庭用水中病原微生物产生的最实用和最低成本的技术之一。印第安部落早在 2000 多年前就使用这种方法净化饮用水。1877 年，Downes 和 Blunt 两人第一次对阳光的杀菌作用进行了实验研究。Arara 等人于 1980 年首次证明了该方法的可行性。自从 20 世纪 90 年代至今，人们对这种光诱导过程的作用机理进行了深入研究。大部分研究集中在水温、太阳能量、污染水浊度、溶解氧、微生物性质等关键参数的作用上，其他研究则是这些技术在不同地点的实际应用和野外试验。

Passage 21 Solar Energy

Since its beginning life has relied upon the sun to sustain a good climate on Earth. Now, the use of solar energy is viewed as a great, non-polluting means of energy. Solar energy is already being used as a source of heat. The principle that a black surface exposed to the sun will absorb solar energy is the basis of several million domestic hot water heaters used in many countries. A more advanced system could be applied to home heating and cooling. If solar energy someday replaces oil and coal, we will enjoy more power, cleaner air, and better health.

Word Bank

1. rely [rɪ'laɪ] *v.* 依靠,信赖
2. sustain [sə'steɪn] *v.* 维持;保持
3. be viewed as... 被看作是……
4. means [miːnz] *n.* 方法;途径
5. non-polluting *adj.* 无污染的
   前缀 non- 置于形容词和名词前构成形容词,指不具有某一特质或表示"非""无"。
   non-alcoholic *adj.* 不含酒精的
   non-stop *adj.* 不停的

6. expose to 暴露于

7. domestic [dəˈmestɪk] *adj.* 国内的；家庭的；驯养的

8. replace [rɪˈpleɪs] *vt.* 代替；替换；把……放回原位；（用……）替换

Exercises

I. Fill in the blanks with the words given below. Change the form where necessary.

| relies | replace | sustain | be viewed as |
| domestic | expose to | dissolve | non-polluting |

1. Because of the slump in _____ demand, production has stopped.
2. Do not throw batteries into fire or _____ high temperature.
3. The problem is how to get enough food to _____ life.
4. Are there any _____ of getting there?
5. She _____ on her parents for tuition.
6. He will be difficult to _____ when he leaves.

II. Comprehension of the passage.

Choose the best answer to each of the following questions or answer the question.

1. Solar energy is _____.
   A. non-polluting
   B. inexpensive to use
   C. potentially dangerous
   D. banned in Japan

2. It is possible that solar energy _____.
   A. may replace oil and coal
   B. may create air and health problems
   C. could blacken the earth's surface
   D. could replace domestic hot water heaters

3. The rays of the sun _____.
   A. cannot be applied to home heating
   B. cannot be used for home cooling
   C. are easily absorbed by a black surface

D. are used for large city lighting

4. List three advantages of using solar energy.

_____
_____
_____

Reference translation

太阳能

从有生命的那一刻起,生命就依靠太阳来维持地球上的良好气候。现在,太阳能被认为是一种非常好且无污染的能源。太阳能已经被用作一种热能。利用暴露在阳光下的黑色外层可以吸收太阳能的原理,很多国家的数百万家庭使用了太阳能热水器。更先进的太阳能系统可应用于家庭取暖和制冷。如果有一天太阳能取代石油和煤,我们将能享有更多的电力、更清洁的空气和更健康的身体。

Passage 22　Happy Bees

Eating sweet food makes people happy. And that feeling leads us to make optimistic choices in ambiguous situations, such as gambling—just like the bees. Other experiments showed that the insects weren't merely more excited or active because of the sugar hit; instead, the treat induced a positive feeling that affected their behavior in other situations. The scientists also ended the bees' optimistic behaviors by giving them a dopamine inhibitor, which blocks the brain's reward center. Optimism is part of human emotions, and finding it in bees makes the scientists themselves optimistic that the feeling has a long evolutionary history and is likely to be found in many animals.

Word bank

1. ambiguous [æmˈbɪgjuəs] *adj.* 模棱两可；含糊的
2. optimistic [ˌɒptɪˈmɪstɪk] *adj.* 乐观的，乐观主义的
3. gambling [ˈgæmblɪŋ] *n.* 赌博；冒险
4. dopamine [ˈdəʊpəmiːn] *n.* 多巴胺（神经细胞产生的一种作用于其他细胞的化学物质）
5. inhibitor [ɪnˈhɪbɪtə(r)] *n.* 抑制剂；抗老化剂
6. block [blɒk] *v.* 阻塞；阻挡　*n.* 块；街区

7. evolutionary [ˌiːvə'luːʃənri] *adj.* 进化的；演变的；逐渐发展的

8. be likely to 可能

9. dose [dəʊs] *n.* 剂量；药量

Exercises

I. Fill in the blanks with the words given below. Change the form where necessary.

| gambling | be likely to | ambiguous | evolutionary |
| dopamine | optimistic | inhibitor | block |

1. Life has its own _____ process.
2. The fires _____ permanently deforest the land.
3. He lost a lot of money in _____.
4. After today's heavy snow, many roads are _____.
5. His answer was _____.
6. She is not _____ about the future.

II. Comprehension of the passage.

Choose the best answer to each of the following questions.

1. The scientists end the bees' optimistic behaviors by _____.
   A. feeding them with sugar
   B. giving them a dopamine inhibitor
   C. tying them up
   D. hypnosis

2. According to the passage, those people who like eating sweet food might are more likely to be _____.
   A. fat
   B. optimistic
   C. excited
   D. active

3. Which of the following statement is true according to the passage? _____
   A. Insects become more excited or active only because of the sugar hit.
   B. Dopamine inhibitor can make bees more active.
   C. The optimistic feeling can be founded in many animals.
   D. Sweet food can keep healthy.

4. The finding of the study can be adapted to _____.
   A. food research   B. body language
   C. biological emotions   D. bees' activity

Reference translation

快乐的蜜蜂

吃甜食会让人有一种快感。这种感觉会诱导我们在模棱两可的情况下去做出一些乐观的选择——比如说沉迷赌博，就像蜜蜂追逐蜜源一样。其他研究表明，这些昆虫不仅仅是因为糖分的刺激而更兴奋或活跃；相反，糖分能诱导出一种积极的感觉从而影响着他们在其他情况下的行为。科学家们通过给他们一种多巴胺抑制剂来结束蜜蜂乐观的行为，它能阻断大脑的反馈中心。乐观是人类情感的一部分，在蜜蜂身上发现这种情感，使科学家们乐观地认为这种快乐感经历了长期的进化过程，而且在许多动物身上都可能找到。

Passage 23　Newly Discovered State of Memory

Memory researchers have shown light into cognitive limbo. A new memory—the name of someone you've just met, for example—is held for seconds in so-called working memory, as your brain's neurons continue to fire. If the person is important to you, the name will over a few days enter your long-term memory. A research team shows that memories can be resurrected from this limbo. Their observations point to a new form of working memory, which they dub prioritized long-term memory, which exists without elevated neural activity. The study doesn't address how synapses or other neuronal features can hold this second level of working memory, or how much information it can store. And this new memory state could have a range of memory-related neurological conditions such as amnesia, epilepsy and schizophrenia.

Word bank

1. cognitive ['kɒgnətɪv] *adj.* 认知的；认识的
2. limbo ['lɪmbəʊ] *n.* 被忽略或遗忘的地方或状态；过渡状态；中间地带
3. neuron ['njʊərɒn] *n.* 神经元；神经细胞
   neuronal [n'jʊərənəl] *adj.* 神经元的
   neurological [ˌnjʊərə'lɒdʒɪkl] *adj.* 神经学的；神经病学的
4. resurrect [ˌrezə'rekt] *v.* 重新应用；使复兴　resurrection [ˌrezə'rekʃn] *n.* 复苏；复兴
5. dub [dʌb] *v.* 授予称号；起绰号；配音
6. prioritize [praɪ'ɒrətaɪz] *v.* 优先处理；按重要性排列；划分优先顺序
7. elevate ['elɪveɪt] *v.* 提高；提升；举起；鼓舞
8. synapse ['saɪnæps] *n.* [生]（神经元的）突触
9. amnesia [æm'niːziə] *n.* [U] 健忘；遗忘（症）
10. epilepsy ['epɪlepsi] *n.* [医]癫痫症
11. schizophrenia [ˌskɪtsə'friːniə] *n.* [医]精神分裂症；矛盾

Exercises

I. Fill in the blanks with the words given below. Change the form where necessary.

elevate	prioritize	ambiguous	resurrection
synapse	limbo	amnesia	cognitive

1. Make lists of what to do and _____ your tasks.
2. This is a _____ of an old story from the mid-70s.
3. Symptoms of too much pressure include insomnia, _____, unexplainable headache and sudden depression.
4. Emotional stress can _____ blood pressure.
5. He is reading a book on _____ psychology.
6. The negotiations have been in _____ since mid-December.

II. Comprehension of the passage.

Choose the best answer to each of the following questions.

1. Which of the following is true? _____
   A. This study solved the synaptic problem.
   B. This study cannot know how much information it can store.
   C. This new memory state could have many uses.
   D. People will be dub prioritized short-term memory.
2. What kind of disease this new memory status might lead to? _____
   A. Split personality.　　　　B. Multiple personality disorder.
   C. Melancholia.　　　　　　D. Irritable sign.
3. Which of the following statement is NOT true according to the passage? _____
   A. Memory can be resurrected from this limbo.
   B. Memory won't be store long time.
   C. The new memory will be lead to neurological conditions.
   D. Memory is held for seconds in so-called working memory.
4. Working memory probably means_____.
   A. a state between long-term memory and short-term memory
   B. long-term memory
   C. short-term memory
   D. memory loss

Reference translation

关于记忆的新发现

记忆研究者已经揭示了人的认知遗忘状态。一个新的记忆点,例如你刚刚遇到的某人的名字,在所谓的工作记忆中保存数秒,因为你大脑的神经元还要继续记忆其他信息。如果这个人对你很重要,那么这个人的名字会进入你的长期记忆中保留多天。一个研究小组表明,记忆可以从这种遗忘中恢复。他们通过观察指出了一种新的工作记忆形式,他们称之为记忆优先长期记忆,这些记忆并没有增加神经元活动性。该研究并没有说明神经结点或其他神经元特征如何能够保持第二级工作记忆,或者可以存储多少信息。这种新的记忆状态可能导致一系列与记忆相关的神经系统疾病,如健忘症、癫痫和精神分裂症。

Passage 24　Advances in Intelligent Vehicles

The book about intelligent vehicles, provides us with up-to-date research results and cutting-edge technologies in the area of intelligent vehicles and transportation systems. It presents recent advances in intelligent vehicle technologies that enhance the safety, reliability, and performance of vehicles and vehicular networks and systems. Topics covered include virtual and staged testing scenarios, collision avoidance, human factors, and modeling techniques.

The series in intelligent systems publishes titles that cover state-of-the-art knowledge and the latest advances in research and development in intelligent systems. Its scope includes theoretical studies, design methods, and real-world implementations and applications.

Word bank

1. cutting-edge ['kʌtɪŋ'edʒ] *adj.* 前沿的；尖端的
2. transportation [ˌtrænspɔː'teɪʃn] *n.* 运送，运输；运输系统；运输工具
3. reliability [rɪˌlaɪə'bɪlətɪ] *n.* 可靠；可信赖
4. vehicular [və'hɪkjələ(r)] *adj.* 车载式的；车辆的

5. scenario [sə'nɑːrɪəʊ] *n.* 设想；预测；（电影的）剧情
6. state-of-the-art [s'teɪt'ɒvðə'ɑːt] *adj.* 使用最先进技术的；体现最高水平的
7. theoretical [ˌθɪə'retɪkl] *adj.* 理论的；假设的；空论的
8. implementation [ˌɪmplɪmen'teɪʃn] *n.* 履行；实施

Exercises

I. Fill in the blanks with the words given below. Change the form where necessary.

| vehicular | scenario | state-of-the-art | transportation |
| reliability | theoretical | cutting-edge | implementation |

1. This is certainly a _____ risk but in practice there is seldom a problem.
2. The fruit will not bear _____ to any great distance.
3. Car buyers are more interested in safety and _____ than speed.
4. We use only the most _____ technology.
5. There is no _____ access.
6. The worst-case _____ would be for the factory to be closed down.

II. Comprehension of the passage.

Choose the best answer to each of the following questions.

1. This report might well be seen _____.
    A. on entertainment newspapers
    B. on science website
    C. in casual magazine
    D. in inspirational book
2. What do we learn about this book and intelligent vehicles?_____
    A. The development of intelligent vehicles improves the safety of vehicles.
    B. This book provides a super research area.
    C. This book contains a wealth of topics and provides the top system.
    D. Intelligent vehicle includes theoretical study and method design.
3. The series in intelligent systems publishes titles that cover _____.
    A. state-of-the-art knowledge
    B. the latest advances in research and development in intelligent systems

C. A & B

D. the problems we are facing

4. What is the passage mainly about? _____

A. Intelligent vehicles bring great convenience to people.

B. Introduction to the research results of intelligent vehicles.

C. Intelligent vehicles can be brought to the community.

D. People's views on intelligent vehicles.

Reference translation

智能汽车的发展

关于智能汽车的书为我们提供了智能汽车和交通系统领域的最新研究成果和尖端技术。书中提出了智能汽车技术的最新发展，这些技术提高了车辆和车辆网络及系统的安全性和可靠性。包括虚拟和分阶段测试场景、避免碰撞、人为因素和建模技术。

智能系统系列的出版物涵盖了最先进的知识以及智能系统研究和开发方面的最新进展。其范围包括理论研究、设计方法以及现实世界的实施和应用。

Passage 25　Compulsive Behavior

Researchers have both created and relieved symptoms of obsessive-compulsive disorder (OCD) in genetically modified mice using a technique that turns brain cells on and off with light, known as optogentics. The work, by two separate teams, confirms the neural circuits that contribute to the condition and points to treatment targets. It also provides insight into how quickly compulsive behaviors can develop and how quickly they might be soothed.

Brain scanning in humans with OCD has pointed to two areas—the orbitofrontal cortex, just behind the eyes, and the striatum, a hub in the middle of the rain—as being involved in the condition's characteristic repetitive and compulsive behaviors. But "in people we have no way of testing cause and effect", says a psychiatrist and neuroscientist who led one of the studies. It is not clear, for example, whether abnormal brain activity causes the compulsions, or whether the behavior simply results from the brain trying to hold symptoms at bay by compensating.

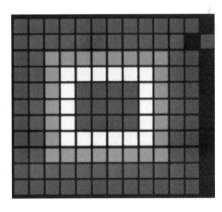

Word bank

1. compulsive [kəm'pʌlsɪv] *adj.* 强迫性的；极有趣的；令人着迷的
2. optogentics [ˌɒptəʊdʒə'netɪks] *n.* 光遗传学
3. confirm [kən'fɜːm] *v.* 证实；确认；批准
4. neural ['njʊərəl] *adj.* 神经的；神经系统的
5. treatment ['triːtmənt] *n.* 治疗；疗法；处理
6. scanning ['skænɪŋ] *n.* 扫描
7. soothe [suːð] *v.* 安慰；抚慰；减轻；缓解

8. orbitofrontal [ˈɔːbɪtəːfrʌntl] *adj.* [医]眶额的

9. cortex [ˈkɔːteks] *n.* 皮质；树皮；果皮

10. striatum [strˈɪeɪtəm] *n.* 纹状体

11. hub [hʌb] *n.* 轮轴；中心，焦点

12. psychiatrist [saɪˈkaɪətrɪst] *n.* 精神病医生；精神病专家

13. neuroscientist [ˈnjʊərəʊsaɪəntɪst] *n.* 神经科学家；神经科学各分支的专家

14. compensate [ˈkɒmpenseɪt] *v.* 补偿；弥补

Exercises

I. Fill in the blanks with the words given below. Change the form where necessary.

| treatment | confirm | cortex | hub |
| compensate | psychiatrist | compulsive | soothe |

1. There are various _____ available for this condition.
2. He was a _____ gambler and often heavily in debt.
3. Nothing can _____ for the loss of a loved one.
4. It is a trade center and transportation _____.
5. His guilty expression _____ my suspicions.
6. Take a warm bath to _____ tense, tired muscles.

II. Comprehension of the passage.

Choose the best answer to each of the following questions.

1. According to the first paragraph, _____ contribute to symptom of obsessive compulsive disorder.

    A. light-sensitive proteins

    B. neural circuits

    C. abnormal brain activities

    D. optical fibers

2. _____ were involved in the condition's characteristic repetitive and compulsive behaviors?

    A. The orbitofrontal cortex and striatum

    B. Eyes and the striatum

C. The orbitofrontal cortex and foreheads

D. The striatum and a hub in the middle of the brain

3. The word "cortex" is closest in meaning to_____.

   A. striatum   B. pericarp

   C. epidermis cell   D. body structure

4. The last word in paragraph 1, "soothe", probably means _____.

   A. comfort   B. detail

   C. persist   D. alleviate

Reference translation

强迫症行为

在转基因鼠的研究中，研究人员使用被称为光遗传学的技术，利用光线控制脑细胞的闭合和开启，能够产生和消除强迫症症状。实验是由两个独立的团队分别进行的，确定了能够影响病情及治疗目标的神经回路。对强迫性行为如何快速产生及如何快速得到缓解提供了新的认识。

对强迫症患者大脑的扫描指向两个区域——在眼睛位置正后方的前额皮层下以及在大脑中央位置的纹状体——它们与重复性和强迫性的行为症状有联系。其中一位从事这项研究的精神病学和神经科学家说，"我们没有办法在人身上测试它的原因和影响"。目前尚不清楚，例如，是否有异常的脑活动导致强迫行为，或者该种行为只是大脑为了抑制症状而做出补偿性行为的一种结果。

Passage 26   Manned Spaceship

The safe return of China's first manned spaceship on October 16, 2003 has made China the third country in the world that has successfully sent man into space following the United States and the Soviet Union. Unlike the unmanned spaceship, the manned spaceship must meet some special technological requirements.

1. Environmental control.

Environmental control aims to regulate the temperature, humidity and pressure in the modules and the spacesuits, absorb the end products of metabolism, handle harmful matters in the modules, and provide oxygen, ventilation and water while disposing of the waste.

2. Safe return.

To ensure the safe return of the astronauts, in addition to the strict temperature control by means of the thermal-protective coating and temperature regulation in the modules, high precision of landing is also very important.

3. High reliability.

All the systems and equipment of the manned spaceship must undergo reliability design, and the key units must have backup systems. The spaceship must pass ground test and simulated flight test under strict control of environments. The design of the spaceship must also allow the astronauts to complete maintenance and replacement in case of equipment failure.

Word bank

1. breakthrough ['breɪkθru:] *n.* 突破；重大进展
2. requirement [rɪ'kwaɪəmənt] *n.* 要求；必要条件
3. regulate ['regjuleɪt] *v.* 管理；约束；调节；控制
4. module ['mɒdju:l] *n.* 模块；组件
5. metabolism [mə'tæbəlɪzəm] *n.* 新陈代谢；代谢作用
6. ventilation [ˌventɪ'leɪʃn] *n.* 通风设备；空气流通
7. dispose [dɪ'spəʊz] *v.* dispose of sb./sth. 去掉；清除；处理；解决
8. thermal-protective coating 热保护层；隔热层
9. undergo [ˌʌndə'gəʊ] *v.* (underwent, undergone) 经历；遭受
10. maintenance ['meɪntənəns] *n.* 维护；维修；维持；保持

Exercises

I. Fill in the blanks with the words given below. Change the form where necessary.

| metabolism | dispose | requirement | regulate |
| breakthrough | undergo | maintenance | modules |

1. You may have to _____ disappointment and failure before experiencing success.
2. All living matter undergoes a process of _____.
3. Patience is a _____ in teaching.
4. These courses cover a twelve-week period and are organized into three four-week _____.
5. They _____ of the city's waste in the sea.
6. An important _____ in negotiations has been achieved.

II. Comprehension of the passage.

Choose the best answer to each of the following questions.
1. From the paragraph 1, we can infer that _____.
   A. China has been the most developed country in the field of space technology in the world

B. the Soviet Union was the first country which sent man into space

C. China has made remarkable achievements in space technology

D. America and the Soviet Union had helped China a lot send man into space

2. The functions of environmental control are in the followings except _____.

    A. handling the harmful material in the modules

    B. addressing the unpredictable emergencies

    C. adjusting the temperature, humidity and pressure in the modules and the spacesuits

    D. allowing the astronauts to complete maintenance and replacement in case of equipment failure

3. Which of the following is true? _____

    A. In order to increase the reliability, strict temperature control is necessary.

    B. There are no obvious differences between the unmanned spaceship and the manned spaceship.

    C. The astronauts don't need the ability to maintain and replace the equipment failure.

    D. The high precision of landing is important to ensure the safe return of the astronauts.

4. In the 3rd point, the reliability design does not include _____.

    A. having backup systems

    B. passing ground test

    C. passing simulated flight test

    D. equipping manned spaceship with emergency equipment

Reference translation

载人飞船

2003年10月16日，中国第一架载人宇宙飞船安全返回地面。这标志着中国成为继美国和苏联之后第三个成功将人类送上太空的国家。与无人飞船相比，载人飞船有一些特殊要求，包括：

1. 环境控制。

环境控制的主要作用是调节舱内和太空服内的温度、湿度和压力，吸收人体新陈代谢产物，控制舱内有害物质，提供氧气、通风、用水并处理废物。

2. 安全返回。

为确保宇航员安全返回，除了通过隔热层和舱内温度调节进行严格的温度控制，较高的落点精度也很重要。

3. 高可靠性。

载人飞船的各系统和设备均要进行可靠性设计，关键部件采用备份系统。飞船须在严格的环境条件下进行地面测试和模拟飞行试验。飞船的设计还要保证宇航员能对有故障的设备进行必要的维修和置换。

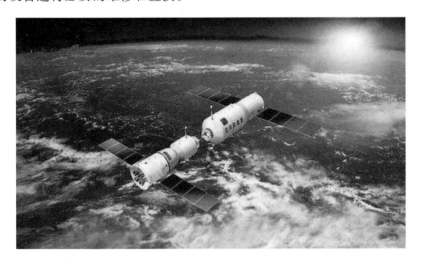

Passage 27　Analysis and Construction of Plastic Degradation

The problem with plastic is they do not easily degrade. They may break down, but only into smaller pieces. The smaller those pieces get, the more places they can go.

Many pieces wind up at sea. Tiny bits of plastic float throughout the world's oceans. They wash up on a remote island. They collect in sea ice thousands of kilometers from the nearest city. They even meld with rock, creating a new material.

Exactly how much plastic is out there remains a mystery. All that missing plastic is worrisome, because the smaller a plastic bit becomes, the more likely it will make its way into a living thing, whether a tiny plankton or an enormous whale.

Word bank

1. degrade [dɪ'greɪd] v. （使）退化；降解；分解
   degradation [ˌdegrə'deɪʃn] n. 降解；分解
2. plastic ['plæstɪk] n. 塑料制品；可塑体；整形 adj. 可塑的；柔软的；有创造力的
3. tiny ['taɪni] adj. 极小的；微小的
4. remote [rɪ'məʊt] adj. 偏远的；遥远的；远程的
5. meld [meld] v. （使）融合；合并；结合
6. remain [rɪ'meɪn] v. 仍然是；保持不变
7. worrisome ['wʌrisəm] adj. 令人担心的
8. plankton ['plæŋktən] n. 浮游生物

Exercises

I. Fill in the blanks with the words given below. Change the form where necessary.

| tiny | remote | worrisome | remain |
| degrades | meld | plastic | plankton |

1. She even had _____ surgery to change the shape of her nose.
2. Train fares are likely to _____ unchanged.
3. The house is _____ from any other buildings.
4. This substance _____ rapidly in the soil.
5. The Martian atmosphere contains only _____ amounts of water.
6. You were all more _____ than a baby.

II. Comprehension of the passage.

Choose the best answer to each of the following questions.

1. Missing small bits of plastic are worrisome because _____.
    A. they can wind up at sea
    B. they can make their way into food chains
    C. they can make their way into a living thing
    D. they can harm human respiratory system
2. What is the problem with plastics? _____
    A. They do not easily degrade.
    B. They do not break down.
    C. They are the major source of soil pollution.
    D. They can meld with rock and create a new material.
3. The tiny bits of plastic can go to _____.
    A. all over the oceans
    B. remote island
    C. ice sea
    D. all of the above
4. In the 3rd paragraph, what does missing plastic refer to _____.
    A. a large part of plastic
    B. tiny plastic

C. bits absorbed by whale
D. bits absorbed by plankton

Reference translation

塑料降解的分析与构建

塑料的问题是它们很难降解。它们可能被分解，但只会粉碎成更小的碎片。这些碎片越小，它们扩散的地方越多。

许多碎片被风吹进大海。微小的塑料漂浮在全世界的海洋中。它们被冲到很远的小岛。它们在离最近的城市数千千米的海冰中聚集。他们甚至可以与岩石融合，形成一种新物质。

究竟有多少塑料是一个谜。所有丢失的塑料都令人担忧，因为塑料碎片越小，它就越容易进入生物体内，无论是微小的浮游生物还是巨大的鲸鱼都能进入。

Passage 28　Atlantic Salmon

Atlantic salmon spawn and spend the juvenile phase in freshwater. After they reach the age and size for smoltification, they typically migrate from river to sea, and perform long-distance marine feeding migrations in the Atlantic Ocean. Atlantic salmon populations have declined in most of the distribution area, and have been lost from all German watersheds due to pollution, migration barriers and habitat destruction. A decline of Atlantic salmon in German rivers likely began with the expansion of watermill technology during the Middle Ages, followed by decreased water quality, habitat degradation and river fragmentation by weirs and dams after the industrial revolution.

Word bank

1. spawn [spɔːn] *v.* 产卵
2. freshwater ['freʃwɔːtə] *n.* 淡水
3. typically ['tɪpɪklɪ] *adv.* 通常
4. migrate [maɪ'greɪt] *v.* 洄游
5. marine [mə'riːn] *adj.* 海洋的
6. decline [dɪ'klaɪn] *v.* 减少；下降
7. distribution [ˌdɪstrɪ'bjuːʃn] *n.* 分布；分发
8. watershed ['wɔːtəʃed] *n.* 水域
9. expansion [ɪk'spænʃən] *n.* 推广；扩张
10. watermill ['wɔːtəmɪl] *n.* 水磨；水车
11. degradation [ˌdegrə'deɪʃ(ə)n] *n.* 退化
12. fragmentation [ˌfrægmen'teɪʃn] *n.* 分裂；破碎
13. weir [wɪə(r)] *n.* 堰　Dujiang Weir 都江堰

14. dam [dæm] *n.* 水坝

15. smoltification [sməʊltifi'keɪʃn] *n.* 一系列生理变化（包括改变身体形状、增加皮肤反射等）

16. Atlantic salmon 大西洋鲑鱼

17. the juvenile phase 幼年期

18. feeding migration 索饵洄游

19. the Atlantic Ocean 大西洋

Exercises

I. Fill in the blanks with the words given below. Change the form where necessary.

| migrate | decline | typically | watershed |
| expansion | marine | spawn | distribution |

1. When I was working in an office, I_____ was sitting down all day.
2. Despite living in different waters, all eels（鳗鱼）_____ in the sea.
3. This is the season when fish _____.
4. It is reported that the temperature will _____ sharply in one or two days.
5. Stream system is a critical and basic unit of _____ management.
6. The above example uses a normal _____.

II. Comprehension of the passage.

Choose the best answer to each of the following questions.

1. According to the passage, Atlantic salmon _____.
   A. were extinct
   B. cannot be killed
   C. migrate from river to sea in order to spawn
   D. undergo many changes before feeding migrations

2. A decline of populations of Atlantic salmon in German rivers probably began in _____.
   A. the Middle Ages
   B. the juvenile phase
   C. the first industrial revolution

D. the second industrial revolution

3. The word "smoltification" is meant to be _____.

   A. the juvenile phase

   B. feeding migrations

   C. spawning migrations

   D. physiological changes

4. Which of the following is not the reason that explains why Atlantic salmon have been lost from all German watersheds according to the short passage? _____

   A. Pollution.

   B. Overfishing.

   C. Migration barriers.

   D. Habitat destruction.

Reference translation

大西洋鲑鱼

大西洋鲑鱼在淡水产卵并度过幼年期，在它们达到一定的年龄和个头后，常从河流洄游到海洋，在大西洋完成长距离的海洋索饵洄游。鲑鱼数量在大部分生活地区都有所减少。由于水质的污染、洄游障碍和栖息地破坏的原因，所有德国水域中的鲑鱼都已消失。德国河流中该鱼的减少可能始于中世纪水磨技术的推广，后因工业革命后水质变差，生息地退化和堤堰、水坝导致的河流细化阻流问题。

Passage 29　Blue Halos

Many wild bees prefer flowers in the violet-blue range—in part because these blossoms tend to produce high volumes of nectar. But it's not easy for plants to produce blue flowers. Instead, many have evolved "blue halos" to allure bees, nanoscale structures on their petals that produce a blue glow when light hits them. The blue halo is created by tiny, irregular striations—usually lined up in parallel fashion. They made their find by using scanning electron microscopy to examine every type of angiosperm—or flowering plant—including grasses, herbaceous plants, shrubs, and trees. The size and spacing of the nanoscale structures vary greatly, yet they all generate a blue or ultraviolet (UV) scattering effect particularly noticeable to bees, which have enhanced photoreceptor activity in the blue-UV parts of the spectrum.

Word bank

1. violet ['vaɪələt] *adj.* 紫色的；紫罗兰色的
2. blossom ['blɒsəm] *v.* 开花；兴旺　*n.* 花；开花期；兴旺期
3. volume ['vɒljuːm] *n.* 量；体积；卷；音量
4. nectar ['nektə(r)] *n.* 花蜜；琼浆玉液
5. allure [ə'lʊə(r)] *v.* 引诱，诱惑；吸引　*n.* 诱惑力
6. nanoscale ['nænoskel] *n.* 纳米级
7. striation [straɪ'eɪʃən] *n.* 条纹；擦痕
8. angiosperm [ˌændʒɪəˌspɜːm] *n.* [植] 被子植物
9. herbaceous [hɜː'beɪʃəs] *n.* 草本的；绿色的；叶状的
10. shrub [ʃrʌb] *n.* 灌木；灌木丛
11. generate ['dʒenəreɪt] *v.* 使形成；发生；生殖
12. scatter ['skætə(r)] *v.* 分散；散开
13. tend to 易于；有……的倾向

Exercises

I. Fill in the blanks with the words given below. Change the form where necessary.

| blossom | allure | violet | generate |
| shrubs | striation | tend to | herbal |

1. The _____ and charm of Paris captivate all who visit there.
2. This is_____ toothpaste.
3. Rain begins to fall and peach trees _____.
4. The light was beginning to drain from a _____ sky.
5. Women _____ live longer than men.
6. I hide myself in bushes and _____.

II. Comprehension of the passage.

Choose the best answer to each of the following questions.

1. Which color attracts bees most according to the passage? _____
   A. Yellow.
   B. Blue.
   C. Green.
   D. White.
2. Which of the following statements is true? _____
   A. Plants attract animals by reflecting blue light.
   B. Bees are only interested in nectar.
   C. The blue halo is created by plants' evolution.
   D. The nanostructure of plants is not very different.
3. This phenomenon is most likely to be used in _____.
   A. tactile sensation
   B. smell
   C. vision
   D. hearing
4. Which of the following description is not true? _____
   A. Produced by tiny, irregular stripe.
   B. Arranged in parallel manner.

C. The size of the nanostructures and spacing varies widely.
D. It will only have a blue scattering effect to attract bees.

Reference translation

蓝色光晕

许多野生蜜蜂喜欢紫色、蓝色的花，部分原因是这些花会产生大量的花蜜。但是，很少有植物开出蓝色的花。于是，许多植物进化出"蓝色光晕"来吸引蜜蜂，花瓣上的纳米结构在光线照射下产生蓝色的光。蓝色的光晕是由微小的、不规则的条纹形成的，通常以平行的方式排列。人们利用电子显微镜扫描每一种被子植物或开花植物，包括草、草本植物、灌木和树木。它们的纳米结构的大小和间距差别很大，但为了吸引蜜蜂，它们都会产生蓝色或紫外（UV）散射效应，这增强了光谱中蓝色紫外部分的感光性。

Passage 30　Childhood Tumors

Childhood tumors aren't duplicates of their adult counterparts. They often carry distinct mutations, respond differently to drugs. Only four new drugs for treating cancer in children have received approval from the U.S. Food and Drug Administration in the past 25 years, versus more than 100 drugs for adult cancers. Now, researchers say they've found a new way to fight cancer in children, one that targets a tumor cell's ability to repair its DNA.

Tumors that strike kids have a potential vulnerability. Most produce PGBD5, a protein that can cause mutations in cancer-thwarting genes, as well as in other genes. A cell that makes lots of PGBD5 needs to be able to repair its DNA to survive, possibly because the protein triggers mutations than can kill the cell. Blocking this repair, you should be able to stop the tumor in its tracks.

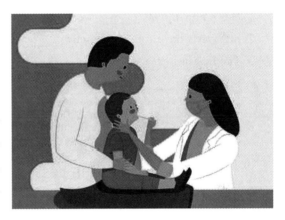

Word bank

1. tumor ['tjuːmə] *n.* 瘤；肿瘤
2. duplicate ['djuːplɪkeɪt] *v.* 重复；复制；*n.* 副本；完全一样的东西；复制品
3. counterpart ['kaʊntəpɑːt] *n.* 配对物；相对物；极相似的人或物
4. mutation [mjuːˈteɪʃn] *n.* 突变；变异；变化；转变
5. versus ['vɜːsəs] *v.* 开花；兴旺　*n.* 花；开花期；兴旺期
6. target ['tɑːɡɪt] *v.* 瞄准；把……目标　*n.* 目标；目的
7. potential [pəˈtenʃl] *adj.* 潜在的；有可能的
8. vulnerability [ˌvʌlnərəˈbɪləti] *n.* 脆弱性；弱点
9. thwart [θwɔːt] *v.* 阻挠；挫败　cancer-thwarting *adj.* 阻止癌症的

10. trigger ['trɪɡə(r)] *v.* 引发；触发

11. block [blɒk] *v.* 阻止；阻塞；限制 *n.* 街区；块；阻碍物

12. track [træk] *n.* 轨道；踪迹；小路

Exercises

I. Fill in the blanks with the words given below. Change the form where necessary.

| versus | trigger | mutation | duplicate |
| potential | track | targeted | counterpart |

1. Our company has _____ women as our primary customers.
2. Scientists have found a genetic _____.
3. The boy has great _____.
4. Stress may _____ these illnesses.
5. He let himself in with a _____ key.
6. The Foreign Secretary telephoned his Indian _____ to protest.

II. Comprehension of the passage.

Choose the best answer to each of the following questions.

1. What is the meaning of "mutations"? _____
   A. Reaction.　　　　　　　　B. Sudden change.
   C. Gene.　　　　　　　　　　D. Substance.
2. PGBD5 _____.
   A. is a protein
   B. can cause mutations in genes
   C. can not kill the cell
   D. All of the above.
3. What is the difference between children cancers and adult cancers? _____
   A. Children cancers are stronger than adult cancers.
   B. Adult cancers have more types.
   C. Children cancers are harder to cure.
   D. Here are more drugs for adult cancers.

4. _____ can stop the tumor in its tracks in children.

   A. Having more drugs

   B. Blocking a tumor cell's ability to repair its DNA

   C. Chemotherapy

   D. Doctors

Reference translation

儿童肿瘤

儿童肿瘤并不是成人肿瘤的复本。它们通常携带不同的突变，对药物有不同的反应。在过去的 25 年里，只有四种治疗儿童癌症的新药获得了美国食品和药物管理局的批准，而治疗成人癌症药物则超过一百种。现在，研究人员说，他们找到了一种新的方法来对抗儿童癌症，一种以阻止肿瘤细胞复制其 DNA 为目标的方法。

危害儿童的肿瘤有潜在的弱点。大多数肿瘤产生的 PGBD5 是一种蛋白质，它可以导致癌症和其他基因的突变。产生大量 PGBD5 的细胞需要复制其 DNA 来生存，这可能是因为蛋白质引发了突变而不是杀死了癌细胞。阻止癌细胞的复制，就能够阻止肿瘤的发生。

Passage 31　Do Animals Recognize Themselves?

Do animals really know who they are? Experts have been puzzling over this intriguing question for decades, and the responses vary depending on whom you ask and how he or she defines self-awareness. The legendary naturalist Charles Darwin believed that humans are not the only self-aware beings. Actually, being self-aware does not necessarily translate into a sense of "I-ness". When an animal sees its reflection, for instance, it may not understand "That's me!" in the same way as a human would, but it may know that its body is its own. Although there is no easy answer to the question of self-awareness in animals, this line of inquiry leads to some fascinating insights into who animals are, what they know and how they feel. And I believe that an animal's awareness of its body and property ultimately equates to a sense of self.

Word bank

1. intriguing [ɪnˈtriːgɪŋ] *adj.* 有趣的；迷人的
2. vary [ˈveəri] *v.* 变化；变异；改变
3. legendary [ˈledʒəndri] *adj.* 传说的；传奇的
4. naturalist [ˈnætʃrəlɪst] *n.* 自然主义者；博物学家
5. reflection [rɪˈflekʃn] *n.* 反映；思考；反射
6. inquiry [ɪnˈkwaɪəri] *n.* 探究；调查；询问
7. fascinating [ˈfæsɪneɪtɪŋ] *adj.* 迷人的，有极大吸引力的
8. insight [ˈɪnsaɪt] *n.* 顿悟；领悟；洞察力
9. ultimately [ˈʌltɪmətli] *adv.* 最后；根本；基本上
10. equate to　相当于；等于

Exercises

I. Fill in the blanks with the words given below. Change the form where necessary.

| reflection | ultimately | intriguing | insight |
| vary | naturalist | fascinating | equate to |

1. The committee _____ adopted his suggestions.
2. Guilin is the most _____ place I have ever been to.
3. People _____ very much in their ideas.
4. Your clothes are often a _____ of your personality.
5. These discoveries raise _____ questions.
6. He is a writer of great _____.

II. Comprehension of the passage.

Choose the best answer to each of the following questions.

1. What conclusion can you get according to the passage? _____
    A. Humans are the only self-aware beings.
    B. The author's view is different from Charles Darwin's.
    C. When an animal sees its reflection, it can understand "That's me!".
    D. The author thinks an animal's consciousness of its body and property in the end is the same as self consciousness.

2. The question "Do animals know who they are?" _____.
    A. has just been raised
    B. leads to some fascinating insights into who animals are, what they know and how they feel
    C. has simple and certain answer
    D. Charles Darwin didn't think about it

3. The author arranges details according to _____.
    A. time order
    B. comparison and contrast
    C. cause and effect
    D. simple listing

4. Which of the following sentences shows the author's opinion? _____

   A. Humans are not the only self-aware beings.
   B. I believe that an animal's awareness of its body and property ultimately equates to a sense of self.
   C. Being self-aware does not necessarily translate into a sense of "I-ness".
   D. There is no easy answer to the question of self-awareness in animals

Reference translation

动物认识自己吗？

动物真的知道它们是谁吗？几十年来，专家们一直对这个有趣的问题感到困惑，回答的不同取决于你问的是谁以及他或她是如何认定自我意识的。著名的博物学家查尔斯·达尔文认为，人类不是唯一有自我意识的生物。事实上，自我意识并不一定转化为某种意义上的"真我"。例如，当一个动物看到它的影子，它可能不能像人类一样意识到"这是我"，但它可能知道它的身体是它自己的。尽管动物的自我意识问题没有一个简单的答案，但这一系列的研究却使我们对动物、它们所知道的以及它们的感受有了一些有趣的见解。我相信动物对它的身体和属性的意识最终等同于自我意识。

Passage 32　Glass Could Hold Ancient Fossils

If you're looking for signs of past life on Earth, some of the evidence is obvious. "Finding a trilobite fossil is a no-brainer, we all can understand that one, or a dinosaur bone." said, a geologist.

Less obvious signs of ancient life can be found in glass. Specifically, impact glass, which forms when asteroids slam into the planet, rapidly heats and melts the rocks around them. Impact glass on the Earth can preserve biological material in a kind of capsule, a time capsule if you will.

For instance, scientists have found impact glass containing ancient plant matter, and other chemical signatures of life.

Some geologists have now spotted the same sort of impact glass on Mars, using the spectrometer on the Mars Reconnaissance Orbiter. Which means Martian glass might also hold evidence of life. If the Red Planet ever harbored life, that is.

Word bank

1. trilobite ['traɪləʊbaɪt] *n.* [古生] 三叶虫
2. no-brainer ['nəʊ'breɪnə] *n.* 非常容易的问题；容易做出的决定
3. asteroid ['æstərɔɪd] *n.* [天]小行星；海星
4. slam [slæm] *v.* 碰撞；砰然关上
slam into/against sb./sth. 重重地撞上
5. preserve [prɪ'zɜːv] *v.* 保护；维护；保留
6. capsule ['kæpsjuːl] *n.* 胶囊；航天舱；（植物的）荚
7. signature ['sɪɡnətʃə(r)] *n.* 明显特征，识别标志；签名，署名
8. the Mars Reconnaissance Orbiter 火星勘测轨道飞行器

9. harbor ['hɑ:bə] *v.* 包含；藏有；心怀 *n.* 港湾；海港

Exercises

I. Fill in the blanks with the words given below. Change the form where necessary.

| capsule | signature | trilobite | preserve |
| no-brainer | harbor | fascinating | asteroid |

1. It's not easy to _____ fresh fruits in summer.
2. The doctor advised me to take a _____ every morning.
3. Bright colors are her _____.
4. This is a _____ solution.
5. The ship was permitted to tie up in Boston _____.
6. I heard the door _____ behind her.

II. Comprehension of the passage.

Choose the best answer to each of the following questions or answer the questions.

1. Which of the following is true about impact glass? _____
   A. The meteorites carry it to earth.
   B. Asteroids slam into the planet, rapidly heating and melting the rocks around them.
   C. People have faked it.
   D. Not mentioned.
2. Some of impact glass called time capsule because _____.
   A. it looks like a capsule
   B. it has a long history
   C. it traveled through time
   D. impact glass on the Earth can preserve biological material in a kind of capsule
3. According to this passage, how do you prove that Mars has life? _____
   A. Land on the Mars, and observe.
   B. Observe whether Martian glass hold evidence of life.
   C. Search for life by remote-controlled rover.
   D. Look for the impact glass.
4. Underline the sentence which shows the author's opinion.

Reference translation

玻璃可以保存古化石

如果你正在寻找过去地球上的生命迹象,有一些证据是显而易见的。"我们都知道,发现一个三叶虫化石或者一个恐龙化石是很容易的",一位地质学家这样说。

玻璃中可以发现不太明显的古代生命迹象。具体是指当小行星撞击行星时形成的冲击玻璃迅速加热并融化它们周围的岩石。在地球上的冲击玻璃可以将生物物质保存在一种"胶囊"中,你可以称之为"时间胶囊"。

例如,科学家已经在冲击玻璃里发现了古代植物物质和其他生命的化学特征。

一些地质学家利用火星轨道勘测飞行器上的光谱仪,在火星上发现了这种冲击玻璃。这意味着火星的玻璃也能证明生命的存在。如果这颗红色星球曾经有生命存在,那么这就是证明。

Passage 33 Laughter

One of the best ways to understand a people is to know what makes them laugh. In humor life is redefined and accepted. Irony and satire provide much keener insights into a group's collective psyche and values than do years of research. It has always been a great disappointment to Indian people that the humorous side of Indian life has not been mentioned by professed experts on Indian affairs. Rather, the image of the granite-faced redskin has been perpetuated by American mythology.

Word Bank

1. laughter ['lɑːftə(r)] *n.* 笑，笑声

2. irony ['aɪrəni] *n.* 反话，讽刺

3. satire ['sætaɪə(r)] *n.* [U] 讽刺，讽刺文学

4. keen [kiːn] *adj.* 尖刻的；敏锐的；强烈的；深切的

5. insight into 洞察

6. psyche ['saɪki] *n.* 心智；灵魂

7. professed [prə'fest] *adj.* 公开声称的，伪称的

8. granite-faced *adj.* 面无表情的

9. redskin ['redskɪn] *n.* 红皮肤人；〈贬〉印第安人

10. perpetuate [pə'petʃueɪt] *vt.* 使永存，使不朽

11. mythology [mɪ'θɒlədʒi] *n.* 神话

12. Indian ['ɪndɪən] *adj.* 印度的，印度人的，印第安人的
    n. 印度人，印第安人，印第安语

Exercises

I. Fill in the blanks with the words given below. Change the form where necessary.

| professed | irony | satire | fabricate |
| laughter | keen | commercially | perpetuate |

1. Competition is very _____.
2. The new building will _____ its founder's great love of learning.
3. He _____ to be content with the arrangement.
4. There is a note of _____ in his voice.
5. The audience roared with _____.
6. The novel is a _____ on American politics.

II. Comprehension of the passage.

Choose the best answer to each of the following questions or answer the questions.

1. The author's point of view is _____.
   A. questionable   B. unconvincing
   C. amusing   D. revealing
2. The author objects to the research done by the experts on Indian affairs because _____.
   A. it overlooks the human element
   B. it is deliberately misleading
   C. it ignores the scientific method
   D. it lacks a sense of humor
3. The author would probably agree with which of the following? _____
   A. A sense of humor is a saving grace.
   B. Laugh and the world laughs with you.
   C. Variety is the spice of life.
   D. Patience overcomes the stubborn will.
4. On the line below write the adjective/noun combination which indicates the author's judgment of those responsible for the myth.

   _____

Reference translation

笑　声

　　了解一个人最好的方法之一就是看什么事可以逗乐他们。通过幽默，生活可以被重新定义和接受。根据多年的研究，反语和讽刺更能让你洞察一个群体的心理和价值观。令印弟安人感到非常失望的是，印弟安人生活中幽默的一面从未被研究印弟安人的专家提及过。相反，美国神话中描述的印弟安人永远是红皮肤、面无表情的形象。

Passage 34　Explores the Martian Upper Atmosphere

Coupling between the lower and upper atmosphere, combined with loss of gas from the upper atmosphere to space, likely contributed to the thin, cold, dry atmosphere of modern Mars. To help understand ongoing ion loss to space, the Mars Atmosphere and Volatile Evolution (MAVEN) spacecraft made comprehensive measurements of the Mars upper atmosphere, ionosphere, and interactions with the sun and solar wind during an interplanetary coronal mass ejection impact in March 2015. Responses include changes in the bow shock and magnetic sheath, formation of widespread diffuse aurora, and enhancement of pick-up ions. Observations and models both show an enhancement in escape rate of ions to space during the event. Ion loss during solar events early in Mars history may have been a major contributor to the long-term evolution of the Mars atmosphere.

Word bank

1. couple ['kʌpl] *v.* 连接；结合 *n.* 对；双；配偶
2. contribute [kən'trɪbjuːt] *v.* ~ (to sth.) 是……的原因之一；促成；促使
3. ongoing ['ɒngəʊɪŋ] *adj.* 不间断的；持续进行的；前进的
4. ion ['aɪən] *n.* [物]离子
5. comprehensive [ˌkɒmprɪ'hensɪv] *adj.* 综合的；广泛的；有理解力的
6. array [ə'reɪ] *n.* 数列；阵列；大量
7. compelling [kəm'pelɪŋ] *adj.* 引人注目的；强制的；激发兴趣的
8. measurement ['meʒəmənt] *n.* [U] 测量；度量 [C] 尺寸；长度；数量
9. ionosphere [aɪ'ɒnəsfɪə(r)] *n.* 电离层
10. interaction [ˌɪntər'ækʃn] *n.* 互动；合作；互相影响
11. interplanetary [ˌɪntə'plænɪtri] *adj.* 行星之间的；行星际的
12. coronal [kə'rəʊnəl] *adj.* 冠的；花冠的 *n.* 冠，花冠，冠状物
13. ejection [ɪ'dʒekʃn] *n.* 射出；喷出
14. bow shock 弓形波
15. sheath [ʃiːθ] *n.* 鞘；护套 magnetic sheath 磁鞘
16. diffuse [dɪ'fjuːs] *adj.* 弥漫的；扩散的；漫射的；不清楚的；冗长的
17. aurora [ɔː'rɔːrə] *n.* 北极光；曙光；朝霞

Exercises

I. Fill in the blanks with the words given below. Change the form where necessary.

| measurement | sheath | diffuse | compelling |
| comprehensive | contribute to | ongoing | interaction |

1. Various factors _____ his failure.
2. They were often _____ to work eleven or twelve hours a day.
3. The police investigation is _____.
4. All the products are labeled with _____ instructions.
5. A _____ book is very tiresome to read.
6. The _____ are extraordinarily accurate.

II. Comprehension of the passage.

Choose the best answer to each of the following questions.

1. Which of the following is not the reason for the formation of the Martian atmosphere? _____

   A. The relationship between the lower and upper atmosphere.

   B. Ion loss in early solar activity.

   C. Loss of gas in the upper atmosphere.

   D. Nature of the atmosphere.

2. Which of the following is not part of the Mars survey? _____

   A. The lower atmosphere of Mars.

   B. Ionosphere.

   C. The interaction between the sun and the solar wind.

   D. The formation of the solar wind.

3. What do you know about atmosphere of modern Mars? _____

   A. Mars has a thin, proper, dry atmosphere.

   B. Scientists have carried out a whole test of the Martian atmosphere, the ionosphere and the interaction with the sun and the solar wind.

   C. The reaction includes the changes of the bow shock and the magnetic sheath, the formation of the widely diffused aurora and the constant absorption of ions.

   D. During the event, the observation and model show the decrease of the ion escape rate.

4. It can be inferred from this passage that _____.

   A. humans can live on Mars

   B. we can improve the atmosphere of Mars by influencing the reaction of the coronal mass

   C. Maven can detect solar activity

   D. the test results show that the ion escape rate of the coronal mass is increased

Reference translation

探索火星大气层

火星形成了稀薄、寒冷而且干燥的大气层，可能是因为它融合了低层大气和高层大气以及损失了高空大气中的气体。为了帮助我们了解丢失于空间的火星离子、火星大气的性质和它的挥发性演变情况，2015 年 3 月，在星际的日冕物质进行抛射时，

MAVEN 航天器全面测量了火星的上层大气、电离层以及太阳和太阳风的相互作用。通过测量发现，日冕物质改变了它的弓激波和磁鞘反应，增加了吸收离子的反应，而且还形成了广泛而弥漫的极光反应。另外，通过观测它的模型发现，它的离子逃逸率在增加。由此可得，火星在历史早期时的太阳活动中失去了离子，这可能是长期影响它大气层的主要原因。

Passage 35　Overweight and Obesity

Overweight and obesity are defined as abnormal or excessive fat accumulation that may impair health. Body mass index (BMI) is a simple index of weight-for-height that is commonly used to classify overweight and obesity in adults. It is defined as a person's weight in kilograms divided by the square of his height in meters (kg/m^2). Definitions by age are as following:

Adults (Above 19)
• overweight is a BMI greater than or equal to 25;
• obesity is a BMI greater than or equal to 30.

Children (Under 5)
• overweight is weight-for-height greater than 2 standard deviations above WHO Child Growth Standards median;
• obesity is weight-for-height greater than 3 standard deviations above the WHO Child Growth Standards median.

Children (Aged between 5~19)
• overweight is BMI-for-age greater than 1 standard deviation above the WHO Growth Reference median;
• obesity is greater than 2 standard deviations above the WHO Growth Reference median.

Word bank

1. obesity [əʊ'biːsətɪ] *n.* 肥胖症；肥胖
2. excessive [ɪk'sesɪv] *adj.* 不间断的；持续进行的；前进的
3. abnormal [æb'nɔːml] *adj.* 反常的；异常的；不规则的
4. accumulation [ə͵kjuːmjə'leɪʃn] *n.* 累积；堆积物
5. impair [ɪm'peə(r)] *v.* 损害；削弱
6. index ['ɪndeks] *n.* 指数；索引
7. be defined as 被定义为；被称为
　　definition [͵defɪ'nɪʃn] *n.* 定义；规定

8. deviation [ˌdiːviˈeɪʃn] *n.* 偏差；背离
9. median [ˈmiːdiən] *n.* 中值；中位数

Exercises

I. Fill in the blanks with the words given below. Change the form where necessary.

| abnormal | excessive | impair | median |
| accumulation | definition | obesity | index |

1. Eating fruits can help reduce _____ of fat.
2. Even one drink can _____ driving performance.
3. I was amazed at his _____ behavior.
4. The Dow Jones _____ fell 10 points this morning.
5. _____ work has drained his energy.
6. What's your _____ of happiness?

II. Comprehension of the passage.

Choose the best answer to each of the following questions.

1. Which figure of BMI can describe that a man aged 23 is overweight? _____
    A. 22.   B. 23.
    C. 24.   D. 25.

2. Which parameter is introduced into the BMI? _____
    A. Weight.
    B. Heart rate.
    C. Vital capacity.
    D. Constellation.

3. What's the BMI of a man aged 26 whose weight and height are 50kg and 1.7m respectively? _____
    A. 17.1.   B. 17.2.
    C. 17.3.   D. 17.4.

4. What does "standard" mean in the passage? _____
   A. Criterion. B. Form.
   C. Content. D. Essence.

Reference translation

超重与肥胖

超重和肥胖被定义为会损害健康的异常或过度的脂肪堆积。身体质量指数（BMI）是一个简单的身高体重指数，常用于判断成人是否超重和肥胖。它的计算方法是人的体重除以其身高的平方（公斤/平方米）。根据年龄的不同，定义如下：

成人（19岁以上）
- 超重是 BMI 大于或等于 25；
- 肥胖是 BMI 大于或等于 30。

儿童（5岁以下）
- 超重是 BMI 大于该年龄段世卫组织生长参考中位数的 2 个标准差；
- 肥胖是 BMI 大于该年龄段世卫组织生长参考中位数的 3 个标准差。

青少年（5~19岁）
- 超重是 BMI 大于该年龄段世卫组织生长参考中位数的 1 个标准差；
- 肥胖是 BMI 大于该年龄段世卫组织生长参考中位数的 2 个标准差。

Passage 36   About Soil

Planting more lentils, chickpeas or other crops known as pulses will improve the health of the world's soils. That information comes from the United Nations Food and Agriculture Organization (FAO). The health of soils around the world has reached critical levels, according to the U. N. agency. About a third of the world's soils are damaged because of erosion, pollution, cities expanding and other issues.

The main damage to soil comes from erosion — the break down and loss of the topsoil by wind, rain and repeated use of machinery. The FAO report said the world is now losing soil 10 to 20 times faster than it is replacing it. Nature takes between 100 and 1,000 years to produce 1 centimeter of soil.

Healthy soil can absorb, or take in, heavy rainfall. If the soil is very firm because of overuse by agricultural machinery, or walking, the rain is not absorbed.

Word bank

1. chickpea ['tʃɪkpiː] *n.* 鹰嘴豆（浅棕色的硬圆豆，可烹食）
2. lentil ['lentl] *n.* 小扁豆；小扁豆植株
3. pulse [pʌls] *n.* 脉搏 (*pl.* pulses) 豆子，豆果果实
4. critical ['krɪtɪkl] *adj.* 关键的；至关重要的；批评的
5. issue ['ɪʃuː] *n.* 问题；担忧；发行物
6. erosion [ɪ'rəʊʒn] *n.* 侵蚀；磨蚀；不规则的
7. expand [ɪk'spænd] *v.* 扩张；扩展
8. topsoil ['tɒpsɔɪl] *n.* 表土层；耕作层
9. replace [rɪ'pleɪs] *v.* 代替；替换
10. overuse [ˌəʊvə'juːz] *n.* 过度使用

Exercises

I. Fill in the blanks with the words given below. Change the form where necessary.

| expand | replace | erosion | issue |
| critical | topsoil | pulses | overuse |

1. Her vocabulary _____ through reading.

2. Who do you suppose will _____ her on the show?

3. Your decision is _____ to our future.

4. Experts say the largest source of resistant bacteria is the _____ and misuse of antibiotics among people.

5. Money is not an _____.

6. As a result of continuous water _____, in the course of time a huge cave was formed.

II. Comprehension of the passage.

Choose the best answer to each of the following questions.

1. What's the main idea of this passage? _____
   A. Planting more lentils, chickpeas in the whole world.
   B. The health of soils around the world has reached critical levels.
   C. Planting chickpeas and lentils can improve the health of soils.
   D. Growing cities and growing food needs.

2. According to this passage, how is some soil destroyed? _____
   A. It is damaged by natural behavior.
   B. Because of some factors of human.
   C. It caused some trouble for the nature.
   D. Soils are damaged because of erosion, pollution, cities expanding and other issues.

3. What's the function of healthy soil? _____
   A. It has no function at all.
   B. It can absorb, or take in, heavy rainfall.
   C. It can protect the plants and our need surroundings.
   D. We will be facing poverty, and we will have more issues of food security.

4. Which of the following is not true? _____

   A. The main damage to soil comes from erosion — the break down and loss of the topsoil by wind, rain and repeated use of machinery.

   B. The FAO report said the world is now losing soil 10 to 20 times faster than it is replacing it.

   C. The soil washes away and may cause flooding.

   D. We will compromise our future.

Reference translation

关于土壤

联合国粮食及农业组织（粮农组织）提供的信息表明，多种植小扁豆、鹰嘴豆或其他豆类作物将改善全球土壤的质量。该机构称，世界各地的土质情况已经达到了临界水平，由于侵蚀、污染、城市扩张和其他问题，约三分之一的世界土壤遭到破坏。

土壤的破坏主要来自侵蚀——风、雨和反复的机械耕种，造成表土的破坏和水土流失。联合国粮农组织的报告称，世界土壤的损失速度比土壤生成要快 10 倍至 20 倍。大自然需要 100 至 1000 年才能生成 1 厘米的土壤。

良好的土壤能吸收纳降水。如果由于农业机械过度使用或碾压使土壤板结，就无法吸收雨水。

Passage 37 Solids and Liquids

What do shoes, paper, and cheese all have in common? They are all solids. Solids are things that have a shape of their own. They do not flow like liquids do. Computers, trees, and soccer balls are also solids.

Liquids do not keep their shape. A liquid can be poured into a container and will take the container's shape. Some examples of liquids are water and milk.

Solids and liquids have something in common. They are both states of matter. Matter is everywhere. It is anything that takes up space and has mass. Mass is a measure of how much matter is in an object. All objects are made of matter.

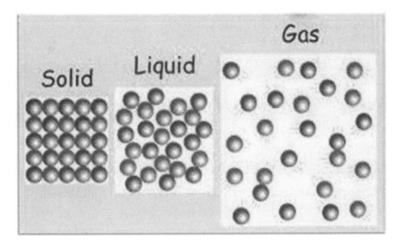

Word bank

1. have sth. in common 有共同之处；有共同点
2. solid ['sɒlɪd] *n.* 固体 *adj.* 固体的；实心的
3. liquid ['lɪkwɪd] *n.* 液体；流音 *adj.* 液体的；清澈的；明亮的
4. flow [fləʊ] *v.* 流；流出
5. pour into 注入；倒入
6. container [kən'teɪnə(r)] *n.* 容器；箱
7. mass [mæs] *n.* [U]质量；[C]团；块；大量

Exercises

I. Fill in the blanks with the words given below. Change the form where necessary.

liquid	flow	common	solid
mass	cheese	container	pour into

1. Keep what's left in a covered _____ in the fridge.
2. The baby is not yet on _____.
3. Photons(光子) have no _____ — they are weightless.
4. Thousands of tons of sewage _____ the Ganges(恒河) every day.
5. Solids turn to _____ at certain temperatures.
6. A stream _____ gently down into the valley.

II. Comprehension of the passage.

Choose the best answer to each of the following questions.

1. What are solids? _____
    A. Things that have a shape of their own.
    B. Water and milk.
    C. Things that do not keep their shape.
    D. Things that flow like liquids.
2. According to this passage, what can be concluded about the shape of water and milk? _____
    A. Water and milk have a shape of their own.
    B. Water and milk do not flow.
    C. Water and milk do not keep their shape.
    D. Water and milk are not liquids.
3. Solids and liquids have something in common. What is it? _____
    A. Solids and liquids are alike in some way.
    B. Solids take up more space than liquids.
    C. Solids and liquids both have a shape on their own.
    D. They are all made of matter.
4. What's the main idea of this article? _____
    A. A liquid that is poured into a container will take the container's shape.

B. Solids and liquids are different kinds of matter.

C. Mass is a measure of how much matter is in an object.

D. Some examples of liquids are water and milk.

Reference translation

固体和液体

鞋子、纸和奶酪有什么共同点？它们都是固体。固体是有自己形状的东西。它们不像液体那样可以流动。电脑、树木和足球也是固体。

液体不能保持固定的形状。液体可以倒进容器中，其形状就会变成容器的形状。比如水和牛奶是液体。

固体和液体有共同之处，它们都是物质的状态。物质无处不在，是任何占一定的空间并具有质量的东西。质量是衡量物体中有多少物质的测量尺度。所有物体都是由物质构成的。

Passage 38　World's Largest Solar Furnace

　　A simple magnifying glass, focusing the sun's rays, can scorch a piece of wood or set a scrap of paper on fire. Solar radiation can also be concentrated on a much larger scale. It can burn a hole through thick steel plate, for example, or simulate the thermal shock of a nuclear blast. The world's largest solar furnace, set up by French scientists high in the Pyrenees, is a complex of nearly 20,000 mirrors. It can concentrate enough sunlight to create temperatures in excess of 6000 degrees Fahrenheit.

　　The furnace's appearance is as spectacular as its power. Its glittering eight-story-high reflector towers over very old houses. For the furnace to operate, these small mirrors must be adjusted so that their light will meet exactly at a focal point 59 feet in front of the giant reflector.

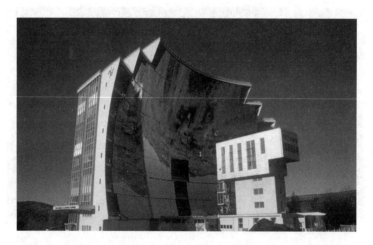

Word Bank

1. furnace [ˈfɜːns] *n.* 熔炉；火炉
2. magnify [ˈmægnɪfaɪ] *v.* 放大；夸大
3. scorch [skɔːtʃ] *v.* 烧焦；烤焦
4. scrap [skræp] *n.* 废料；小片
5. plate [pleɪt] *n.* 金属板；盘子
6. thermal [ˈθɜːml] *adj.* 热的；保暖的

7. Fahrenheit ['færənhaɪt] *adj.* 华氏的；华氏温度计的；*n.* 华氏温度计；华氏温标

8. spectacular [spek'tækjələ(r)] *adj.* 惊人的；壮观的

9. glittering ['glɪtərɪŋ] *adj.* 闪闪发光的；辉煌的

Exercises

I. Fill in the blanks with the words given below. Change the form where necessary.

| furnace | scrap | Fahrenheit | magnify |
| spectacular | plate | scorch | glittering |

1. There was a _____ sunrise yesterday.

2. He has a _____ career ahead of him.

3. People often use a loudspeaker to _____ the voice.

4. Keep away from the _____.

5. By mid-morning, the temperature was already above 100 degrees _____.

6. The leaves are inclined to _____ in hot sunshine.

II. Comprehension of the passage.

Choose the best answer to each of the following questions or answer the questions.

1. The world's largest solar furnace _____.

   A. is made of thousands of magnifying glasses

   B. is a large and impressive structure

   C. blends well with its surroundings

   D. operates even in cloudy weather

2. We can assume that the French solar furnace _____.

   A. produces tons of steel every day

   B. has been operating for many years

   C. is still experimental

   D. will be used for nuclear tests

3. The large reflector of the solar furnace _____.

   A. collects and intensifies sunlight

   B. reflects the sun's rays back into space

   C. bounces sunshine to other nearby reflectors

D. protects local residents from solar radiation

4. Underline the sentence which proves that setting the giant reflector is a delicate operation.

Reference translation

世界上最大的太阳能炉

用一个简单的放大镜聚焦阳光,可以烧焦一块木头或点燃一张纸。太阳辐射光也可以聚集在一起。例如,它可以在厚钢板烧出一个洞,或者模拟核爆炸的热冲击。世界上最大的太阳能炉在法国比利牛斯山,是由近两万个镜面组成的复杂装置。它可以聚集足够的阳光以得到6000华氏度以上的温度。

太阳能炉的外表和它的能量一样很壮观。它闪闪发光的八层高的反射塔耸立在一栋旧房屋上。为了使太阳能炉运行,必须调整小反射镜面,使它们的反射光线正好位于所构成的巨大反射镜前面59英尺的焦点上。

Passage 39　Self-healing Polymeric Material

Few things in this day and age are as disheartening as seeing your smart phone fall to the ground and witnessing its screen crack or shatter. What would you do if that happens?

Thanks to researchers who have developed a self-healing polymeric material that could one day be used on a smart phone screen, allowing it to repair damage in less than 24 hours.

The researchers developed the material from a stretchable polymer and an ionic salt, which are joined together by a special kind of bond called an ion-dipole interaction. This bond is a force between ions and polar molecules in which they are attracted to each other in order to fix the damage. This material is capable not only of stretching itself up to 50 times its regular size, but also repairing itself in less than 24 hours—even if it is completely torn in half.

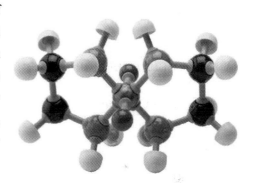

Word bank

1. dishearten [dɪs'hɑːtn] *v.* 使沮丧；使灰心
2. polymer ['pɒlɪmə(r)] *n.* 聚合物
   polymeric [ˌpɒlɪ'merɪk] *adj.* 聚合的
3. molecule ['mɒlɪkjuːl] *n.* 分子；微小颗粒
4. witness ['wɪtnəs] *v.* 当场看到；目击；见证
5. heal [hiːl] *v.* （使）康复，复原；治愈
   self-healing ['selfhiːl] *adj.* 自愈的
6. stretch [stretʃ] *v.* 拉长；撑大；伸展　stretchable *adj.* 有弹性的
7. ionic [aɪ'ɒnɪk] *adj.* 离子的
   ion-dipole interaction　离子偶极作用
8. be capable of　有能力；有才能
9. up to　多达

Exercises

I. Fill in the blanks with the words given below. Change the form where necessary.

| witness | crack | ionic | heal |
| molecule | dishearten | stretch | capable |

1. The ice _____ as I stepped onto it.
2. India has _____ many political changes in recent years.
3. Don't let this defeat _____ you.
4. You are _____ of better work than this.
5. Is there any way of _____ shoes?
6. Minor cuts can usually be left uncovered to _____ by themselves.

II. Comprehension of the passage.

Choose the best answer to each of the following questions.

1. When your smart phone fell to the ground and its screen cracked or shattered, you might well _____.
    A. buy a new phone
    B. break your heart
    C. let it heal itself
    D. throw it away
2. A stretchable polymer and an ionic salt are joined together by _____.
    A. a covalent bond which is strong but difficult to reform once broken
    B. a noncovalent bond which is weaker but reform far more easily
    C. a bond called an ion-dipole interaction
    D. a kind of glue
3. What is the greatest advantage of this material? _____
    A. Its cheap price.
    B. It is stretchable.
    C. It is not easy to reform.
    D. It can fix itself.
4. How many times can the size of the extensible polymer be stretched to? _____
    A. 24.   B. 50.   C. 48.   D. 36.

Reference translation

自修复高分子材料

在当今这时代，几乎没有什么事情比看着你的智能手机掉到地上屏幕摔坏更令人沮丧的了。如果发生这种情况，你会怎么办？

感谢这些研究人员，他们已经开发出一种可以在智能手机屏幕上使用的自愈聚合材料，它可以在不到24小时内修复损伤。

研究人员利用一种可拉伸的聚合物和离子盐，通过一种称为离子偶极相互作用的特殊键结合在一起的方式而开发了这种材料。这种结合是利用离子和极性分子之间的力，它们互相吸引以修复损伤。这种材料不仅能在其被拉伸到原来50倍的情况修复，而且能在不到24小时内自身得到修复，即使它被完全拉分成两半时也可以做到。

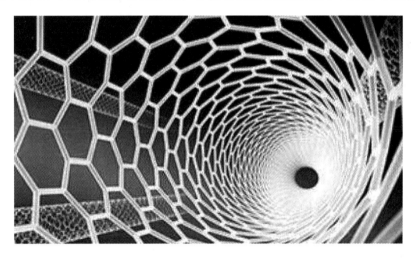

Passage 40 Unified Theory of Evolution

Today, this is the precise definition of environmental epigenetics: the molecular factors that regulate how DNA functions and what genes are turned on or off, independent of the DNA sequence itself. Epigenetics involves a number of molecular processes that can dramatically influence the activity of the genome without altering the sequence of the DNA in the genes themselves.

One of the most common such processes is "DNA methylation", in which molecular components called methyl groups (made of methane) attach to DNA, turning genes on or off, and regulating the level of gene expression. Environmental factors such as temperature or emotional stress have been shown to alter DNA methylation, and these changes can be permanently programmed and inherited over generations–a process known as epigenetic transgenerational inheritance.

Word bank

1. unified ['juːnɪfaɪd] *adj.* 统一的；一元化的
2. epigenetics [ˌepɪdʒ'netɪks] *n.* 实验胚胎学；表观遗传学
3. regulate ['regjuleɪt] *v.* 调节；调整；控制
4. sequence ['siːkwəns] *n.* 顺序；序列
5. dramatically [drə'mætɪklɪ] *adv.* 戏剧性的；引人入胜的
6. genome ['dʒiːnəʊm] *n.* 基因组；染色体组
7. alter ['ɔːltə(r)] *v.* 改变；更改
8. methylation [meθɪ'leɪʃn] *n.* 甲基化作用

9. permanently ['pɜ:mənətlɪ] *adv.* 永久的；长期不变的
10. inherit [ɪn'herɪt] *v.* 继承；经遗传获得（品质、身体特征等）
    inheritance [ɪn'herɪtəns] *n.* 继承；遗传

Exercises

I. Fill in the blanks with the words given below. Change the form where necessary.

permanent	inherit	alter	genome
regulate	unify	sequence	dramatically

1. A _____ contains all of the genetic information about an organism.
2. The First Emperor of Qin _____ China in 221 B.C.
3. The cost of living has increased _____.
4. We need laws to _____ market behavior.
5. Heavy drinking can cause _____ damage to the brain.
6. He had _____ so much I scarcely recognized him.

II. Comprehension of the passage.

Choose the best answer to each of the following questions.

1. Which factors can regulate DNA and genes according to epigenetics? _____
   A. Temperature.
   B. Molecular factors.
   C. Emotional stress.
   D. Methyl groups.

2. Which of the following statements is true about how molecular factors affecting heredity? _____
   A. Molecular processes influence the activity of the genome without changing the sequence of the DNA.
   B. Molecular processes influence the activity of the genome by changing the sequence of the DNA.
   C. Molecular factors influence the activity of the genome by combine with the DNA.
   D. Molecular factors influence the activity of the genome because they are part of genes.

3. _____ can be changed by environmental factors according to the passage?

   A. Sequence of the DNA

   B. Molecular factors

   C. The level of gene expression

   D. DNA methylation

4. Which description is true about "DNA methylation"? _____

   A. Environmental factors such as temperature or emotional stress have been shown to alter DNA but not DNA methylation.

   B. Methyl groups turning genes on or off, and regulating the level of gene expression.

   C. Methyl groups are attached to DNA, and then play a role.

   D. DNA methylation can be permanently programmed and inherited between finite generations.

Reference translation

统一进化论

如今，环境表观遗传学的确切定义为：可以调节 DNA 功能以及激活或影响基因的分子因素，与 DNA 序列本身无关。表观遗传学涉及许多分子过程，它们可以显著影响基因组的活性而不改变基因本身的 DNA 序列。

其中最常见的一个过程是"DNA 甲基化"，在这个过程中，称为甲基（由甲烷构成）的分子附着于 DNA，打开或关闭基因，并调节基因表达水平。环境因素如温度或情绪压力已被证明可以改变 DNA 甲基化，而这些变化可以永久进行并代代相传，这一过程被称为表观遗传代际遗传。

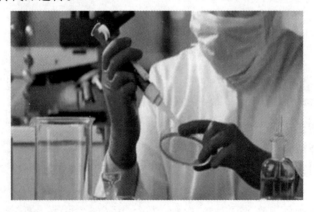

Passage 41 Creativity and Childhood Experiences

Performing artists who were exposed to abuse, neglect or a dysfunctional family as a child might experience their creative process more intensely, according to a new long-term study that has found a link between the two. Psychologists studied 234 professional performers, looking for a reason why mental health disorders are so common in the performing arts. The notion that artists and performing artists suffered more pathology, including bipolar disorder, troubled us, psychologist told Psypost. The study examined 83 actors, directors, and designers; 129 dancers; and 20 musicians and opera singers. These study participants filled out self-report surveys pertaining to childhood adversity, sense of shame, creative experiences, proneness to fantasies, anxiety, and level of engagement in an activity.

Word bank

1. creativity [ˌkriːeɪˈtɪvətɪ] n. 创造力；创造性
2. abuse [əˈbjuːz] n. 滥用；虐待；辱骂 v. 滥用；辱骂
3. dysfunctional [dɪsˈfʌŋkʃənl] adj. 功能失调的；功能障碍的
4. intensely [ɪnˈtenslɪ] adv. 强烈地；剧烈地
5. pathology [pəˈθɒlədʒi] n. 病理（学）；异常状态

6. bipolar [ˌbaɪˈpəʊlə(r)] *adj.* 有两极的，双极的，两极世界的；(心理学) 双相型障碍的；躁狂抑郁性精神病的
7. participant [pɑːˈtɪsɪpənt] *n.* 参与者；参加者
8. pertain [pəˈteɪn] *v.* 关于；与……相关
9. proneness [ˈprəʊnnəs] *n.* 倾向

Exercises

I. Fill in the blanks with the words given below. Change the form where necessary.

| participant | dysfunctional | participant | co-author |
| creativity | intensely | pertain | pathology |

1. The fire flamed _____.
2. He has been an active _____ in the discussion.
3. _____ is used to describe relationships or behaviors which are different from what is considered to be normal.
4. What she did was an _____ of her authority.
5. Seeds will not without water. Scientific research often involves _____ and inventions.
6. My remark _____ to your earlier comments.

II. Comprehension of the passage.

Choose the best answer to each of the following questions.

1. According to the passage, creativity has something to do with _____.
    A. household income
    B. education
    C. childhood experiences
    D. individual endowment
2. "Performing artists" are probably _____.
    A. directors
    B. actors
    C. dancers
    D. doctors

3. Which of the following does not include in the self-report? _____

   A. Superiority complex.

   B. Sense of shame.

   C. Fantasies.

   D. Anxiety.

4. We can see from the passage that _____.

   A. psychologists are looking for a reason why mental health disorders are so common in children

   B. performing artists suffered more pathology

   C. Paula Thomson is one of the performing artists being studied

   D. the study is fruitful

Reference translation

创造力与童年经历

一项长期研究发现，表演艺术家的创造力和童年经历存在某种联系，童年的不幸，如被虐待、被忽视、家庭不和睦，会使其更富创造力。心理学家调查了234名专业表演艺术家，寻找表演艺术家心理健康问题频发的原因。心理学家在网站说："艺术家和表演家更容易患上躁郁症等心理疾病，这让我们感到困扰。"该研究调查了83名演员、导演和设计师；129名舞者；20名音乐家和歌剧演唱家。这些参与者填写了有关童年不幸、羞耻感、创作体验、幻想倾向、焦虑和活动参与度的自述报告。

Passage 42　The Risk of Excessive Dietary Salt

A diet rich in salt is linked to an increased risk of cerebrovascular diseases and dementia, but it remains unclear how dietary salt harms the brain.

We report that, in mice, excess dietary salt suppresses resting cerebral blood flow and endothelial function, leading to cognitive impairment. The effect depends on expansion of TH17 cells in the small intestine, resulting in a marked increase in plasma interleukin-17(IL-17). Circulating IL-17, in turn, promotes endothelial dysfunction and cognitive impairment by the Rho kinase–dependent inhibitory phosphorylation of endothelial nitric oxide synthase and reduced nitric oxide production in cerebral endothelial cells.

The findings reveal a new gut–brain axis linking dietary habits to cognitive impairment through a gut-initiated adaptive immune response compromising brain function via circulating IL-17. Thus, the TH17 cell-IL-17 pathway is a putative target to counter the deleterious brain effects induced by dietary salt and other diseases associated with TH17 polarization.

Word bank

1. cerebrovascular [ˌserəbrəʊ'væskjələ] *adj.* 脑血管的
2. dementia [dɪ'menʃə] *n.* （医）痴呆
3. suppress [sə'pres] *v.* 抑制（感情等）；压制；阻止……的生长（或发展）
4. cerebral ['serəbrəl] *adj.* 大脑的；理智的
5. endothelial [ˌendə'θiːlɪəl] *adj.* （医）内皮的
6. intestine [ɪn'testɪn] *n.* （解）肠
7. plasma ['plæzmə] *n.* 血浆；原生质，细胞质；乳清
8. kinase ['kɪneɪs] *n.* 激酶；致活酶
9. inhibitory [ɪn'hɪbɪtərɪ] *adj.* 禁止的；抑制的

10. phosphorylation [fɒsfɒrɪ'leɪʃən] *n.* 磷酸化（作用）

11. nitric ['naɪtrɪk] *adj.* 氮的；含氮的

12. oxide ['ɒksaɪd] *n.* 氧化物

13. synthase ['sɪnθeɪs] *n.* 合酶

14. putative ['pjuːtətɪv] *adj.* 一般认定的；推定的；假定存在的

15. polarization [ˌpəʊləraɪ'zeɪʃn] *n.* 极化；产生极性；（光）偏振

Exercises

I. Fill in the blanks with the words given below. Change the form where necessary.

intestine	dementia	plasma	impairment
polarization	suppress	synthase	inhibitory

1. Wine or sugary drinks are _____ for digestion.

2. This vitamin is absorbed through the walls of the small _____.

3. _____ is the clear liquid part of blood which contains the blood cells.

4. She was unable to _____ her anger.

5. He has a visual _____ in the left eye.

6. _____ is a serious illness of the mind.

II. Comprehension of the passage.

Choose the best answer to each of the following questions.

1. Scientists did the experiment with _____.

   A. cats

   B. mice

   C. birds

   D. rabbits

2. According to the passage, it is _____ that high salt diet affects the brain.

   A. sure

   B. necessary

   C. unclear

   D. important

3. Which of the following is true about the TH17 cell-IL-17 pathway? _____

A. It's very important for the human body.

B. It's harmful to the human body.

C. It can reduce the harm of high salt food to the human body.

D. It does not really exist.

4. The passage most probably taken from a _____.

A. entertainment magazine

B. diet book

C. medical journal

D. fairy tale

Reference translation

高盐饮食的危害

高盐的饮食与脑血管疾病和痴呆的风险增加有关,但目前还不清楚饮食中的盐如何危害大脑。

据报告,过量的食盐会抑制小鼠的静息脑血流量和内皮功能,导致认知障碍。这种效应依赖于小肠中 TH17 细胞的扩增,导致血浆白细胞介素-17(IL-17)明显增加。循环 IL-17 依次通过一氧化氮合酶的 Rho 激酶依赖性抑制磷酸化,促使内皮功能障碍和认知功能障碍,以及减少脑内皮细胞一氧化氮的产生。

这些发现揭示了一种新的肠—脑轴,通过肠道启动的适应性免疫反应,循环 IL-17 来破坏大脑功能,从而将饮食习惯与认知障碍二者联系起来。因此,TH17 细胞-IL-17 通路是用来对抗由食盐和其他与 TH17 极化相关的疾病引起的有害脑效应的一个设定的目标。

Passage 43 The Discovery of Artemisinin

Artemisinin, also known as Qinghaosu, and its semi-synthetic derivatives are a group of drugs used against Plasmodium falciparum malaria. It was discovered by Tu Youyou, a Chinese scientist, who was awarded half of the 2015 Nobel Prize in Medicine for her discovery. Treatments containing an artemisinin derivative (artemisinin-combination therapies, ACTs) are now standard treatment worldwide for P. falciparum malaria. Artemisinin is isolated from the plant Artemisia annua, sweet wormwood, a herb employed in Chinese traditional medicine. A precursor compound can be produced using genetically engineered yeast.

Artemisinin and its endoperoxides derivatives have been used for the treatment of P. falciparum related infections but low bioavailability, poor pharmacokinetic properties and high cost of the drugs are a major drawback of their use. Therapies that combine artemisinin or its derivatives with some other antimalarial drug are the preferred treatment for malaria and are both effective and well tolerated in patients.

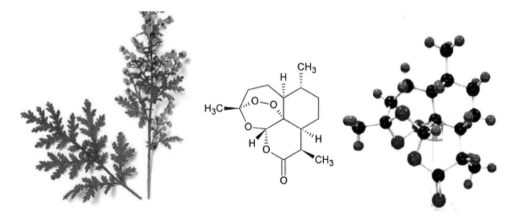

Word bank

1. semi-synthetic ['semɪsɪnθ'etɪk] *adj.* 半合成的
2. derivative [dɪ'rɪvətɪv] *n.* [化] 衍生物；派生物
3. plasmodium falciparum malaria 恶性疟原虫

4. artemisia annua 香青蒿；黄花蒿

5. herb [hɜːb] *n.* 草本植物；药草；香草

6. precursor [priˈkɜːsə(r)] *n.* 前辈；前驱；先兆；初期形式

7. yeast [jiːst] *n.* 酵母（菌）；酵母粉

8. compound [ˈkɒmpaʊnd] *n.* 复合物

9. endoperoxide [endəʊpəˈrɒksaɪd] *n.* 内过氧化物

10. bioavailability [ˌbaɪəʊəˌveɪləˈbɪlɪtɪ] *n.* 生物药效率，生物利用度

11. pharmacokinetic [ˌfɑːməkoʊknetɪk] *n.* [医]药物代谢动力学

12. property [ˈprɒpəti] *n.* 特性；属性；财产

13. drawback [ˈdrɔːbæk] *n.* 缺点；劣势；退税

Exercises

I. Fill in the blanks with the words given below. Change the form where necessary.

compound	yeast	property	precursor
drawback	herb	endoperoxide	derivative

1. This medicinal _____ is hard to come by.

2. The only _____ of this city is its traffic.

3. In chemistry, a _____ is a substance that consists of two or more elements.

4. Real tennis, an ancient _____ of the modern game, originated in the eleventh century.

5. Be careful not to damage other people's _____.

6. Mix the flour and _____ together until you have a sticky dough.

II. Comprehension of the passage.

Choose the best answer to each of the following questions.

1. It was _____ who discovered the artemisinin.

   A. a Chinese scientist

   B. an American scientist

   C. an Indian scientist

   D. a Japanese scientist

2. Artemisinin can be used to treat _____.

   A. systemic lupus erythematous

   B. AIDS

   C. Plasmodium falciparum malaria

   D. lung cancer

3. What are the medicinal disadvantages of artemisinin? _____

   A. Low bioavailability.

   B. Poor pharmacokinetic properties and high cost of the drugs.

   C. A & B.

   D. None of the above.

4. Which of the following is true according to the passage? _____

   A. Tu Youyou is the winner of the 2015 Nobel Prize in Literature for her discovery.

   B. Artemisinin is easy to extract from hundreds of plants.

   C. Plasmodium falciparum malaria is treatable.

   D. Artemisinin is easy to extract from hundreds of animals.

Reference translation

青蒿素的发现

青蒿提取物又称青蒿素，它和它的半合成衍生物都是用于抗恶性疟原虫的药物。青蒿素是中国科学家屠呦呦发现的，她获得了 2015 年诺贝尔医学奖的一半。含有青蒿素衍生物的疗法（青蒿素联合疗法，ACTs）现在是治疗恶性疟原虫疟疾的全球标准治疗方法。青蒿素是从中药青蒿，一种甜蒿中分离得到的。前体化合物可以使用基因工程酵母来生产。

青蒿素及其内过氧化物衍生物已用于治疗恶性疟原虫相关感染，但生物利用度低、药代动力学性质差、药物成本高是其主要缺点。将青蒿素或其衍生物与其他抗疟药物结合是疟疾的首选治疗方法，对患者治疗有效且耐受性良好。

Passage 44　NASA's Plans to Send Humans to Mars

Martian wannabes will have to wait a little bit longer before landing on the enigmatic red planet, according to NASA. The space agency says it's about 25 years away from sending the first human to Mars and has spelled out what it will take to make this space travel plan pan out. Scientists still have to work out major kinks–such as deadly radiation from the cosmos, potential vision loss, and atrophying bones–before sending a human to Mars, known for its unpredictable atmosphere. They're also trying to shave some time off the commute: It could take nine months or so to reach the red planet, which is located about 140 million miles from Earth.

Also, floating in prolonged zero gravity can cause acute medical conditions, such as irreversible changes to blood vessels in the retina, leading to vision degradation, and the skeleton can leach calcium and bone mass. With former astronaut Tom Jones admitting budget constraints will hobble NASA's attempts to send its first human for about 25 years, those of us dreaming of a vacation to Mars dubiously look to SpaceX as our only hope.

Word bank

1. Martian ['mɑːʃn] *n.* 火星人　*adj.* 火星的
2. wannabe ['wɒnəbi] *n.* 超级崇拜者
3. enigmatic [ˌenɪɡ'mætɪk] *adj.* 神秘的；高深莫测的
4. kink [kɪŋk] *n.* 弯；结
5. cosmos ['kɒzmɒs] *n.* 宇宙
6. atrophy ['ætrəfi] *v.* 萎缩；衰退　*n.* 萎缩
7. commute [kə'mjuːt] *n.* 通勤来往；通勤来往的路程　*v.* 通勤

8. irreversible [ˌɪrɪ'vɜːsəbl] *adj.* 不可逆的；无法复原（或挽回）的

9. retina ['retɪnə] *n.* 视网膜

10. degradation [ˌdegrə'deɪʃn] *n.* 恶化；坠落

11. leach [liːtʃ] *v.* 过滤；滤去

12. hobble ['hɒbl] *v.* 阻止；妨碍；蹒跚；跛行

13. dubiously ['djuːbɪəslɪ] *adv.* 可疑地；怀疑地

Exercises

I. Fill in the blanks with the words given below. Change the form where necessary.

| cosmos | kink | enigmatic | atrophy |
| commute | irreversible | dubiously | hobble |

1. I have only a short _____ to work.

2. Our world is but a small part of the _____.

3. Memory can _____ through lack of use.

4. The old man _____ across the road.

5. His only answer was an _____ smile.

6. We need to iron out the _____ in the new system.

II. Comprehension of the passage.

Choose the best answer to each of the following questions.

1. The "wannabes" are most probably people who ____.

   A. are the fans of science fictions

   B. are interested in exploring Mars

   C. lived on Mars

   D. are aliens

2. If possible, it will take the astronauts _____ to commute to Mars.

   A. about twenty-five years

   B. about twelve years

   C. about nine months

   D. about nine years

3. Which of the following medical conditions does not result from floating in

prolonged zero gravity? _____

    A. Irreversible changes to blood vessels in the retina.

    B. Vision degradation.

    C. Loss of calcium and bone mass.

    D. Loss of memory.

4. NASA may settle the financial matter by _____.

    A. loaning money from the state bank

    B. resorting to commercial banks

    C. cooperating with private enterprise

    D. intensifying structural adjustment

Reference translation

美国国家航空航天局计划将人类送上火星

美国国家航空航天局（NASA）称，火星迷们在登上这颗神秘的红色星球前还需要再等上一段时间。该机构表示，将首位人类送上火星还需要大约25年的时间，并阐明了如何才能使这一太空旅行计划成功。科学家们在将人类送上不可预知大气环境的火星之前，还需要解决一些重大问题，比如来自宇宙的致命辐射、潜在的视力丧失和骨骼萎缩。他们还试图减少往返时间：目前需要9个月左右的时间才能到达这颗距离地球1.4亿英里的红色行星。

此外，在失重状态下长时间漂浮会导致急性疾病，比如视网膜血管发生不可逆转的变化导致视力下降，骨骼会流失钙和骨质。前宇航员汤姆·琼斯承认，有限的预算将阻碍美国国家航空航天局，使得他们要花大约25年才能尝试将首位人类送上火星，我们当中那些梦想着去火星度假的人只能试着把SpaceX公司视为唯一的希望了。

Passage 45　New Artificial Nerves

Prosthetics may soon take on a whole new feel. That's because researchers have created a new type of artificial nerve that can sense touch, process information, and communicate with other nerves much like those in our own bodies do. Future versions could add sensors to track changes in texture, position, and different types of pressure, leading to potentially dramatic improvements in how people with artificial limbs—and someday robots—sense and interact with their environments. Modern prosthetics are already impressive: Some allow amputees to control arm movement with just their thoughts; others have pressure sensors in the fingertips that help wearers control their grip without the need to constantly monitor progress with their eyes.

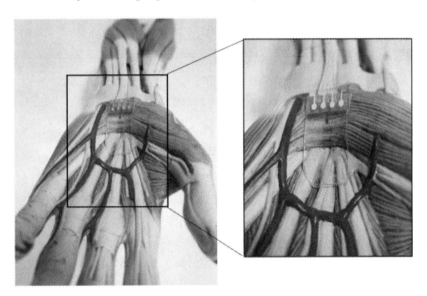

Word bank

1. prosthetics [prɒs'θetɪks] *n.* [*pl.*] 假体；义肢
2. version ['vɜːʃn] *n.* 版本；译文；描述
3. sensor ['sensə(r)] *n.* 传感器；灵敏元件
4. track [træk] *v.* 跟踪；追踪　*n.* 小路；小道

5. texture ['tekstʃə(r)] *n.* 质地；结构
6. limb [lɪm] *n.* 枝干；肢
7. amputee [ˌæmpjuˈtiː] *n.* 被截肢者
8. grip [grɪp] *n.* 握力；紧握

Exercises

I. Fill in the blanks with the words given below. Change the form where necessary.

| version | grip | texture | amputee |
| prosthetics | limbs | sensor | track |

1. For a while, she lost the use of her _____.
2. If you buy a drier, look for one with a _____ which switches off when clothes are dry.
3. Both the _____ and condition of your hair should improve.
4. This is an updated _____ of his book.
5. She tried to get a _____ on the icy rock.
6. We continued _____ the plane on our radar.

II. Comprehension of the passage.

Choose the best answer to each of the following questions.

1. According to the passage, the artificial nerve can _____.
   A. capture touch
   B. handle information
   C. interact with other nerves
   D. assist human body

2. Modern prosthetics are already impressive because _____.
   A. some allow amputees to control arm movement with just their thoughts
   B. others have pressure sensors in the fingertips
   C. the need to constantly monitor progress with their eyes
   D. A&B

3. Which the following is author's attitude about the artificial nerve? _____
   A. Expectant.

B. Indifferent.

C. Confused.

D. Irrelevant.

4. Which sentence indicates the impression of the modern artificial prosthetic? _____

   A. Some allow amputees to control arm movement with just their thoughts.

   B. Others have pressure sensors in the fingertips that help wearers control their grip without the need to constantly monitor progress with their eyes.

   C. Prosthetics may soon take on a whole new feel.

   D. A&B.

Reference translation

新人造神经

假肢很快就会具有全新的感觉。这是因为研究人员发明了一种新型的人工神经，这种神经能够感知触觉、处理信息，并像我们身体里的神经一样与其他神经进行交感。未来的设计可能会增加传感器来跟踪纹理、位置变化和不同类型的压力的变化，这可能会极大地改善人们使用人工肢体的方式。很快，机器人将能够感知环境并与其互动。现代假肢已经让人们印象深刻：一些假肢允许截肢者用他们的思想控制手臂的运动；另一些人的指尖装有压力传感器，可以帮助他们控制其抓握力，而不需要一直用眼盯着。

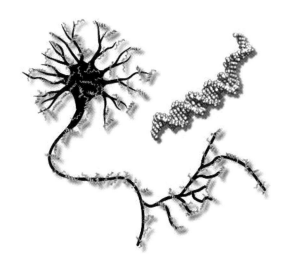

Passage 46　Blood Donation and Cardiovascular Disease

Recently, researchers simulated the working hours of night workers with zebrafish, a model organism. The effects of the jet lag syndrome on animals were found to show signs of red blood cell aging. Aging red blood cells gather on the walls of blood vessels, increasing the risk of clots that can lead to heart disease. This also explains why night-shift workers are 30% more likely to have a heart attack than normal people. In addition, aging red blood cells also reduce their function, especially the ability to transport oxygen in the blood. People who regularly donate blood are similar to zebrafish's response to hypoxia and can trigger a new red, researchers say. The production of cytokines may reduce the risk of cardiovascular disease among nocturnal workers.

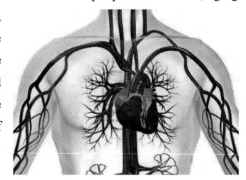

Word bank

1. donation [dəʊ'neɪʃn] *n.* 捐赠；赠送
2. organism ['ɔːɡənɪzəm] *n.* 有机体；生物体
3. jet [dʒet] *n.* 喷气式飞机；喷雾
4. lag [læɡ] *n.* 滞后；落后　*v.* 落后；滞后于
5. syndrome ['sɪndrəʊm] *n.* 综合征；综合症状
6. clot [klɒt] *n.* 凝块　*v.* 凝结；拥塞
7. hypoxia [haɪ'pɒksiə] *n.* 缺氧；低氧
8. trigger ['trɪɡə(r)] *v.* 引发；触发
9. cytokine [ˌsɪtə'kɪn] *n.* 细胞活素
10. cardiovascular [ˌkaːdɪəʊ'væskjələ(r)] *adj.* 心血管的
11. nocturnal [nɒk'tɜːnl] *adj.* 夜的；夜间的

Exercises

I. Fill in the blanks with the words given below. Change the form where necessary.

| trigger | syndrome | clot | organism |
| lag | nocturnal | cytokine | donation |

1. Most bats and owls are _____.
2. The little boy _____ behind his parents.
3. Stress may _____ these illnesses.
4. They made a generous _____ to charity.
5. They removed a _____ from his brain.
6. This _____ is associated with frequent coughing.

II. Comprehension of the passage.

Choose the best answer to each of the following questions.

1. Which is true about red blood cells? _____
   A. The red blood cells were their research subjects.
   B. Red blood cells age.
   C. Red blood cells have nothing to do with heart disease.
   D. Zebrafish doesn't have red blood.

2. What is the regular result of a blood donation? _____
   A. Production of new erythropoietin.
   B. Working longer at night.
   C. Living longer life.
   D. Heart disease.

3. Which of the following is the main idea of the passage? _____
   A. Studies on zebrafish.
   B. Blood transport capacity.
   C. Erythrocyte senescence.
   D. Relationship between blood donation and heart disease.

4. Which of the following factors has something to do heart disease? _____
   A. Over-eating.

B. Working too hard during the day.
C. Working at night shift.
D. Fulminating anoxia.

Reference translation

献血与心血管疾病

　　最近研究人员通过斑马鱼模拟了夜间工作者的工作时间。他们发现这种时差综合征对动物的影响表现为红细胞的衰老。衰老的红细胞会聚集在血管壁上，增加了形成血块从而引发心脏病的概率。这也解释了夜班工作者比正常人的心脏病发作率增加了30%的原因。另外，衰老的红细胞还会降低它们的机能，尤其是在血液中对氧气的运输能力会下降。研究人员称，人有规律的献血类似于斑马鱼在缺氧环境下的反应，能够激发产生新的红细胞，有可能会降低夜间工作人员心血管疾病的发病率。

Passage 47　Superglue

Nowadays, because scientists have created a new kind of glue that can bond hard and soft substances to hydrogels, jello-like materials used in everything from medical devices to soft robots. Previously, researchers in these fields used an ultraviolet light treatment, but it could take up to an hour or more to attach the surfaces together. Now, a team of experimental physicists has invented a new adhesive, made of superglue's main ingredient—cyanoacrylate—plus an organic compound that diffuses into the parts being fused, leading to a tough bond without brittle residue left behind. This nonsolvent delays the hardening of the glue just long enough to let it seep into each layer being pressed together, forming a bond within seconds. The hydrogel bond can hold up to 1 kilogram and stretch up to 2000%.

Word bank

1. superglue ['su:pəglu:] *n.* 强力胶
2. hydrogel ['haɪdrədʒel] *n.* 水凝胶
3. ultraviolet [ˌʌltrə'vaɪələt] *adj.* 紫外线的
4. treatment ['tri:tmənt] *n.* 处理；治理；治疗；疗法
5. cyanoacrylate [ˌsaɪənəʊ'ækrəleɪt] *n.* 氰基丙烯酸盐黏合剂
6. diffuse [dɪ'fju:s] *v.* 弥漫；发散
7. brittle ['brɪtl] *adj.* 易碎的；难以相处的
8. residue ['rezɪdju:] *n.* 剩余物；残留物；残渣
9. nonsolvent [nɒn'sɒlvənt] *n.* 非溶剂

10. seep [siːp] *v.* 渗，渗透

11. downside ['daʊnsaɪd] *n.* 缺点；不利方面

Exercises

I. Fill in the blanks with the words given below. Change the form where necessary.

| brittle | residue | seep | nonsolvent |
| ultraviolet | superglue | downside | treatment |

1. Water _____ from a crack in the pipe.
2. What is the _____ of chocolate for heart heath?
3. The sun's _____ rays are responsible for both tanning and burning.
4. _____ things break easily.
5. Many patients are not getting the medical _____ they need.
6. Always using the same shampoo means that a _____ can build up on the hair.

II. Comprehension of the passage.

Choose the best answer to each of the following questions.

1. _____ is needed to attach the surfaces together before the invention superglue.
    A. Ultraviolet ray
    B. Infrared ray
    C. Sunlight
    D. X-ray

2. In addition to its main ingredient, what kind of substance does the new superglue include? _____
    A. Cyanoacrylate.
    B. An organic compound.
    C. Hydrogels.
    D. Ultraviolet ray.

3. Which of the following is true according to the passage? _____
    A. Superglue is used widely.
    B. It takes only one hour to connect the surface.
    C. The bonded objects can't be stretched.

D. It will not be available in the next few years.

4. Which sentence describes the excellent performance of superglue?_____

   A. This nonsolvent delays the hardening of the glue just long enough to let it seep into each layer being pressed together, forming a bond within seconds.

   B. It won't be on the market for another 3 to 5 years.

   C. It could take up to an hour or more to attach the surfaces together.

   D. Researchers in these fields used an ultraviolet light treatment.

Reference translation

超强力胶水

如今，科学家发明了一种新的胶水。用一种像果冻一样的水凝胶材料，能将软硬物质黏在一起，可用于黏接从医疗设备到软体机器人的所有东西。此前，这些方面的研究人员是利用紫外光进行处理，但这可能需要一个小时或者更多时间才能将表面黏接在一起。现在，一个由实验物理学家组成的团队发明了一种新的黏合剂。它由超强力胶水的主要成分——氰基丙烯酸酯——加上一种会渗入被黏合部件的有机化合物制成，从而产生一种不会留下易碎残渣的韧性黏接剂。这种非溶剂会将胶水的硬化时间延长到刚好使其渗入被压在一起的各层中，从而在几秒钟内形成粘结剂。这种水凝胶黏结剂能承受1千克重量，并且可拉伸至原来的20倍。

Passage 48 The Pavements that Generate Solar Energy

An international team is creating pavements that gather energy from the sun. Solar panels are nothing new but these tiles are designed to lock together in their hundreds to create a solar pavement. They are coated in a tough epoxy resin and will have a scuff and slip-proof finish in a range of colors. But why seek solar power beneath people's feet? The spread of solar energy means roof space will be a diminishing resource. Meanwhile, cities are getting denser as electricity demand rises. So it will become a fact that a new source of energy can be provided to us through these pavements.

Word bank

1. pavement ['peɪvmənt] *n.* 人行道；路面
2. tile [taɪl] *n.* 瓦片；瓷砖
3. epoxy [ɪ'pɒksi] *adj.* 环氧的
4. resin ['rezɪn] *n.* 树脂
5. scuff [skʌf] *v.* 磨损；磨坏

6. slip-proof 防滑

7. diminishing [dɪ'mɪnɪʃɪŋ] *adj.* 逐渐缩小的；逐渐减少的

8. dense [dens] *adj.* 密集的；稠密的

Exercises

I. Fill in the blanks with the words given below. Change the form where necessary.

| tile | scuff | pavement | diminishing |
| resin | dense | synthase | inhibitory |

1. _____ marks from shoes are difficult to remove.

2. Universities are facing grave problems because of _____ resources.

3. The _____ from which the oil is extracted comes from a small, tough tree.

4. Java is a _____ populated island.

5. Tommy's shoes squeaked on the _____ as she walked down the corridor.

6. He was hurrying along the _____.

II. Comprehension of the passage.

Choose the best answer to each of the following questions.

1. The international team is creating the pavements for _____.

    A. backing a international-based project

    B. having fans walking in sunshine

    C. gathering energy from the sun

    D. nothing

2. Why does the author think the solar panels are special? _____

    A. They are coated in a tough epoxy resin.

    B. They will have a scuff and slip-proof finish in a range of colors.

    C. They are designed to lock together in their hundreds to create a solar pavement.

    D. All of above.

3. From the passage we can see that _____.

    A. roof space will be a decreasing resource

    B. solar panels are new to us

    C. the international-based project has won an award

D. people enjoyed the solar energy produced beneath feet

4. People seek solar power beneath feet because _____.

   A. solar panels are nothing new

   B. the spread of solar energy means roof space will be a diminishing resource

   C. cities are getting denser as electricity demand rises

   D. B & C

Reference translation

产生太阳能的人行道

一个国际研究小组正在创造一种可收集太阳能量的人行道。太阳能电池板不是什么新鲜事，但将其和很多瓷砖连在一起创建一个太阳能路面则是一个很独特的想法。路面涂上坚硬的环氧树脂后，这些色彩鲜艳的路面耐磨并可防滑。不过为什么人们会在脚下寻求太阳能呢？太阳能的普及使得在屋顶利用太阳能的空间在不断地减少。与此同时，随着电力需求的增长，城市变得越来越密集。所以，通过这些路面人们将获得一种新的能量来源。

Passage 49   The Role of Embryonic Stem Cells

Scientists already knew that embryonic stem cells can differentiate into any of the body's specialized cell types: bone and brain, lung and liver. They also knew that special groups of cells found in amphibian and fish embryos play an executive role in shaping early developmental structures. These groups, called "organizers", emit molecular signals that direct other cells to grow and develop in specific ways. When an organizer is transplanted from one embryo to another, it spurs its new host to produce a secondary spinal column and central nervous system, complete with spinal cord and brain. Due to the ethical guidelines that limit experimentation on human embryos, however, they did not know if a similar organizer existed in humans.

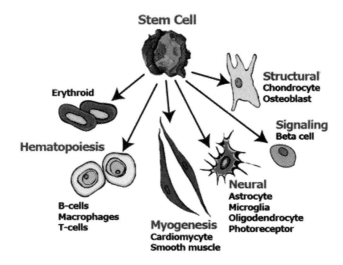

Word bank

1. embryonic [ˌembri'ɒnɪk] *adj.* 胚胎期的；萌芽期的；未成熟的
2. embryo ['embriəʊ] *n.* 胚；胚胎
3. differentiate [ˌdɪfə'renʃieɪt] *v.* 区分；区别；辨别
4. amphibian [æm'fɪbiən] *n.* 两栖动物

5. emit [i'mɪt] *v.* 发出；射出；散发

6. spur [spɜː(r)] *v.* 马刺；刺激；鼓舞

7. spinal ['spaɪnl] *adj.* 脊柱的；脊髓的

8. ethical ['eθɪkl] *adj.* 道德的；伦理的

Exercises

I. Fill in the blanks with the words given below. Change the form where necessary.

| specialize | dementia | spur | amphibian |
| embryonic | ethical | emit | spinal |

1. The frog is an _____, which means it can live on land and in water.

2. Her difficult childhood _____ her on to succeed.

3. The metal container began to _____ a clicking sound.

4. This movement lengthens your spine and tones the _____ nerves.

5. I can't _____ one variety from another.

6. The plan, as yet, only exists in _____ form.

II. Comprehension of the passage.

Choose the best answer to each of the following questions.

1. _____ is capable of evolving into other cells.

   A. Blood cells

   B. Embryonic stem cells

   C. Bone cells

   D. Brain cells

2. What is the function of embryonic stem cells? _____

   A. Differentiation and organization.

   B. Growing up by themselves.

   C. Attacking normal cells.

   D. Anything possible.

3. What will happen when an embryo transplanted to another? _____

   A. It will die.

   B. It will change itself.

C. It will be destroyed by other cells.

D. It will stimulate another one to grow up.

4. Which of the following statement is true according to the passage? _____

A. Embryonic stem cells can differentiate into any of the body's specialized cell types.

B. Emit molecular signals that direct other cells to grow and develop in specific ways.

C. Embryonic stem cells can complete with spinal cord and brain.

D. Due to the ethical guidelines, they did not know if a similar organizer existed in humans.

Reference translation

胚胎干细胞的角色

科学家们已经知道胚胎干细胞可以分化为身体的任何一种特殊细胞类型：骨骼和大脑、肺和肝脏。他们也知道这种特殊细胞在两栖类和鱼类胚胎形成早期发育结构中所起的作用。这些组织被称为"组织者"，用特定的方法发出分子信号，引导其他细胞生长和发育。当一个组织者从一个胚胎移植到另一个胚胎时，它会刺激它的新宿主产生一个次级脊柱和中枢神经系统，以及完整的脊髓和大脑。然而，由于限制人类胚胎实验的伦理准则，他们不知道人类是否存在类似的组织者。

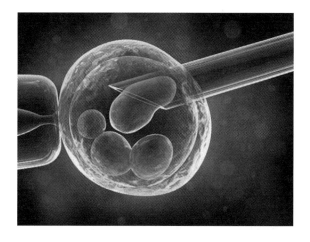

Passage 50　The Secret in the Fingerprint

Before the discovery of DNA profiling in the 1980s, fingerprints were the easiest way to solve serious crimes. It's believed that of the 7 billion or so people on Earth, each one of us has our own unique fingerprints. Fingerprints can indicate lifestyle and environment, eating habits, possible medical problems and even the job of a person. So how can we figure all this out from just a simple fingerprint?

A fingerprint is formed when a finger makes contact with a surface. Most fingerprints are invisible to the naked eye and require a chemical development process in order to make them visible. Stuck between the ridges of a fingerprint, however, are substances that can tell a story about who we are. Things like traces of sweat, blood, and food reveal a lot of information about us—what we've touched, what we've eaten and even what drugs we've taken.

Word bank

1. profile ['prəufaɪl] *v.* 概述；扼要介绍　*n.* 轮廓
2. billion ['bɪljən] *n.* 十亿；数以十亿计
3. invisible [ɪn'vɪzəbl] *adj.* 看不见的；隐形的
4. ridge [rɪdʒ] *n.* 山脊；山脉；背脊
5. trace [treɪs] *n.* 痕迹；踪迹；微量
6. reveal [rɪ'viːl] *v.* 揭露；泄露

Exercises

I. Fill in the blanks with the words given below. Change the form where necessary.

| ridge | reveal | fingerprint | invisible |
| profile | trace | substance | billion |

1. He wanted to be over the line of the _____ before the sun had risen.
2. They have spent _____ on the problem.
3. She has refused to _____ the whereabouts of her daughter.
4. This picture shows the girl in _____.
5. She was _____ in the dusk of the room.
6. I _____ the course of the river on the map.

II. Comprehension of the passage.

Choose the best answer to each of the following questions.

1. What is the main idea of this passage? _____
   A. Fingerprints were the easiest way to solve serious crimes.
   B. It's believed that of the 7 billion or so people on Earth, each one of us has our own unique fingerprints.
   C. Fingerprints are unique. They hide a lot of information.
   D. The formation of fingerprints is due to contact between fingers and objects.
2. We learned from the passage that _____.
   A. all fingerprints are invisible to the naked eye
   B. fingerprints can indicate age
   C. fingerprints are of so much fun
   D. fingerprints can indicate a lot of information of a person
3. The author's attitude about fingerprint is _____.
   A. indifferent
   B. suspicious
   C. negative
   D. positive
4. Which of the following statement is true according to the passage? _____
   A. After the discovery of DNA profiling in the 1980s, fingerprints were the easiest

way to solve serious crimes.
B. Everyone of us has our own unique fingerprints.
C. All of fingerprints are invisible to the naked eye.
D. Fingerprints reveal a lot of information about us–what we've touched, what we've eaten and even what drugs we've taken.

Reference translation

指纹里的秘密

在 20 世纪 80 年代 DNA 分析技术问世之前，指纹是破解重大刑事案件最容易的方法。人们相信，在地球约 70 亿的人口中，每个人的指纹都是独一无二的。指纹可以体现一个人的生活方式、所处环境、饮食习惯、潜在的健康问题甚至职业。所以我们如何从一个指纹中获知这一切呢？

当手指和物体表面接触时，指纹就形成了。大多数的指纹都并非肉眼可见，需要经过化学显影处理后方可现形。而在指纹隆起的纹路之间就隐藏着包含各种信息的物质。汗渍、血液以及食物的痕迹透露了许多相关信息——比如你接触了什么、吃了些什么、甚至是服用了什么药。

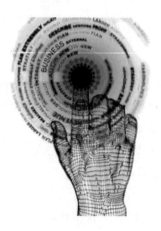

Passage 51 Bees Have Different Taste in Flowers

Male and female bees may look similar, but they have dramatically different dining habits, according to a new study. Despite both needing nectar to survive, they get this nutrient from different flowers—so different, in fact, that males and females might as well belong to separate species. To make the find, researchers observed the foraging habits of 152 species of bees in several flower-rich fields. Then they brought the insects—nearly 19,000 in all—back to the lab and meticulously identified their species and sex. Males and females rarely drank nectar from the same type of flower, the team reports. Using a statistical test the researchers found that male and female bee diets overlap significantly less than would be expected at random.

The preferences probably stem from the distinct physiological and reproductive demands of the two sexes. Males take nectar only for immediate energy, and they typically avoid flowers that produce no nectar. Females—the worker bees—consume nectar, too, but also carry pollen from the fields to their hive. These tasks require females to visit a greater diversity of flowers.

Word bank

1. nectar ['nektə(r)] *n.* 花蜜；琼浆玉液
2. forage ['fɒrɪdʒ] *v.* 搜寻，尤指动物觅食
3. meticulously [mə'tɪkjələsli] *adv.* 非常细致地；无微不至地
4. overlap [ˌəʊvə'læp] *n.* 重叠部分；覆盖物 *v.* 重叠；与……部分相同
5. random ['rændəm] *adj.* 任意的；随机的；胡乱的
6. statistical [stə'tɪstɪkl] *adj.* 统计的；统计学的
7. pollen ['pɒlən] *n.* 花粉
8. diversity [daɪ'vɜːsəti] *n.* 差异；多样化

Exercises

I. Fill in the blanks with the words given below. Change the form where necessary.

forage	statistical	diversity	random
nectar	overlap	meticulously	pollen

1. The upper layer of felt should _____ the lower.
2. On such a hot day, even water was _____.
3. The report contains a great deal of _____ information.
4. We disturbed a wild boar that had been _____ by the roadside.
5. We received several answers, and we picked one at _____.
6. She is always _____ accurate in punctuation and spelling.

II. Comprehension of the passage.

Choose the best answer to each of the following questions.

1. What's the main point of this article? _____
    A. Male and female bees look similar.
    B. Both of male and female bees need honey.
    C. Male and female bees have different eating habits.
    D. Male bees take a lot of time to harvest honey.
2. The researchers chose the flower-rich fields for observation _____.
    A. because of the big size of the fields there
    B. because of the spectacular scenery there
    C. because it's full of flowers
    D. because it's the hometown of the researchers
3. The last sentence suggests that _____.
    A. male and female bees rarely eat the same food
    B. male bees must eat the same food
    C. female bees have to eat the same food
    D. male and female bees usually eat the same food
4. What is the difference between male and female bees according to the passage? _____
    A. Their outward appearance.

B. They must be the different species.

C. They get their own nutrient from different flowers.

D. Female bees need to move honey while male bees don't.

Reference translation

蜜蜂对花有不同的品味

根据一项新的研究，雌蜂和雄蜂可能看起来相似，但它们的就餐习惯却有很大的不同。尽管它们都需要花蜜才能存活下来，但它们从完全不同的花朵中获取所需的营养。事实上，它们的不同之处在于，雄性和雌性可能属于不同的物种。为了取得这一发现，研究人员在几个盛产花卉的地方观察了 152 种蜜蜂的觅食习惯。然后，他们把这些昆虫——总共将近 1.9 万只——带回实验室，仔细地鉴定了它们的种类和性别。研究小组称，雄性和雌性很少吃同一种花的花蜜。通过一项统计测试，研究人员发现，雌雄蜜蜂的饮食重叠程度明显低于随机预期。

这种偏好可能源于两性独特的生理和生殖需求。雄性只采花蜜作为即时能量，它们通常避开不产花蜜的花朵。雌蜂，即工蜂也吃花蜜，但也把花粉从田里带到蜂巢里。这些任务要求雌蜂探访更多种类的花。

Passage 52　Ants and Free Radicals

We humans take medicine when we're sick. As do our primate cousins. Chimps, for example, snack on a bitter African shrub to combat intestinal worms. But the habit extends even to invertebrates. Take fruit flies—which sip alcohol to ward off parasitic wasps. Or wood ants, which line their nests with ant fungal, antibacterial tree sap.

Now researchers report that ants there that have encountered a pathogenic fungus appear to fight the infection by eating foods high in free radicals. Those are molecules with a talent for causing cell damage, in this case, to the cells of the fungus.

The researchers collected some 400 wild ants. They exposed some to the fungus, and left the rest alone. Then they offered up a sort of egg custard—either plain, or laced with free radicals—in the form of hydrogen peroxide. Uninfected ants didn't want anything to do with the radical-rich food. Which makes sense.

An insect immunologist said "Exactly, we don't take painkillers on a daily basis, or we don't take antimicrobial agents on a daily basis. Because that would have really severe side effects on the organism".

Word bank

1. intestinal [ˌɪntes'taɪnl] *adj.* 肠的
2. invertebrate [ɪn'vɜːtɪbrət] *n.* 无脊椎动物
3. ward [wɔːd] *v.* ward off 防止，避免，使防止（危险、疾病、攻击等）
4. parasitic [ˌpærə'sɪtɪk] *adj.* 寄生的
5. wasp [wɑsp] *n.* 黄蜂；胡蜂；易怒的人
6. pathogenic ['pæθə'dʒenɪk] *adj.* 致病的；病原的；发病的
7. custard ['kʌstəd] *n.* 蛋奶糊；蛋羹
8. peroxide [pə'rɒksaɪd] *n.* 过氧化氢；过氧化物 *vt.* 以过氧化氢漂白；以过氧化物处理
9. immunologist [ˌɪmjʊ'nɒlədʒɪst] *n.* 免疫学家

Exercises

I. Fill in the blanks with the words given below. Change the form where necessary.

| shrub | wasp | peroxide | invertebrate |
| pathogenic | custard | ward | intestinal |

1. The fire flamed. She put up her hands to _____ him off.

2. There is a small evergreen _____ on the hillside.

3. He was stung by a _____.

4. Lipid _____ created by fats combined with oxygen tends to build up in the body and create aging.

5. We had plums and _____ for dinner.

6. A _____ organism can cause disease in a person, animal, or plant.

II. Comprehension of the passage.

Choose the best answer to each of the following questions.

1. The ants fight the fungus infection _____.

    A. by eating egg custard

    B. by exercising

    C. by eating plain food

    D. by eating radical-rich food

2. Why can free radicals deal with fungus infections? _____

    A. Because it has the ability to cause cell damage.

    B. Because it has the ability to cause cell recovery.

    C. Because it kills the infected fungus directly.

    D. Because it doesn't play any role.

3. According to the experiment, _____.

    A. infected ants eat common food

    B. uninfected ants eat radical-rich food

    C. infected ants eat radical-rich food

    D. uninfected ants eat a little free radical food

4. The word "combat" in the second line probably means _____.

    A. treat   B. resist

    C. compete   D. enclose

Reference translation

蚂蚁与自由基

我们人类在生病时吃药。我们的灵长类表亲也一样。例如，黑猩猩吃苦涩的非洲灌木来对抗肠道蠕虫。这种习惯甚至延伸到无脊椎动物身上。比如果蝇喝酒以防寄生蜂；还有木蚁，它们的巢穴里有蚂蚁的真菌和抗菌树液。

现在研究人员报告说，遇到致病性真菌的蚂蚁似乎可以通过食用富含自由基的食物来对抗感染。这些分子具有导致细胞损伤的能力，在这种情况下，对真菌的细胞造成伤害。

研究人员收集了大约400只野蚂蚁，把其中一些暴露在真菌中，将另外一些与其他的隔离。然后他们提供了一种蛋羹——一种普通的，另一种含有自由基——以过氧化氢的形式。未受感染的蚂蚁不想和含有自由基的食物扯上任何关系，这是有道理的。

一位昆虫免疫学家说："没错，我们不用每天服用止痛药，也不需要每天服用抗菌剂。因为那样会对机体产生严重的副作用。"

Passage 53　Sleeping

Many scientists have pondered the question of why sleeping gives our brain such a boost. After all, it'd be ideal if we didn't need to sleep at all: shut-eye makes animals vulnerable to predators. They think sleep is important for two main reasons: It helps us repair and restore our organ systems including our muscles, immune systems, and various other hormones. And it plays a crucial role in memory, helping us retain what we learned at work or school for later use.

Getting proper sleep, scientists have found, seems to help our immune systems function best. While our body is resting, immune cells known as T-cells spend that time racing around our bodies. Other immune cells also work better with more sleep. Researchers studied how our bodies respond to vaccines—medicine that targets the immune system—after a full night's rest and after no sleep at all. They found that getting proper sleep the night after a vaccine creates a stronger immune response to the virus a given vaccine is meant to attack.

Word bank

1. ponder ['pɒndə(r)] *v.* 思索 衡量
2. boost [buːst] *n.* 提高；增加； *v.* 增加；促进
3. vulnerable ['vʌlnərəbl] *adj.* （身体上或感情上）脆弱的，易受……伤害的
4. predator ['predətə(r)] *n.* 以掠夺为生的人；食肉动物
5. crucial ['kruːʃl] *adj.* 决定性的；关键性的；极其显要的
6. retain [rɪ'teɪn] *v.* 保持；留在心中
7. hormone ['hɔːməʊn] *n.* 激素；荷尔蒙
8. vaccine ['væksiːn] *n.* 疫苗，痘苗

Exercises

I. Fill in the blanks with the words given below. Change the form where necessary.

predator	vaccine	crucial	retain
ponder	vulnerable	hormone	boost

1. She looked very _____ standing there on her own.
2. The interior of the shop still _____ a nineteenth-century atmosphere.
3. The company is worried about takeovers by various _____.
4. She lay awake all night _____ whether to leave or stay.
5. There is no _____ against HIV infection.
6. Improved consumer confidence is _____ to an economic recovery.

II. Comprehension of the passage.

Choose the best answer to each of the following questions.

1. According to the passage, people need enough time to sleep because of _____.
   A. avoiding predators
   B. keeping healthy
   C. making us feel better
   D. surviving

2. Our immune system needs sleep to _____.
   A. work better
   B. race around our body
   C. attack virus
   D. create stronger immune response

3. According to the passage, enough sleep help us _____.
   A. restore our system
   B. increase ability to recall
   C. keep mind and body healthy
   D. make us energetic

4. Which of the following sentences is true? _____
   A. Immune system needs sleep to work better.
   B. Sleeping is the best way to keep body healthy.

C. Most people suffer from chronic sleep loss.

D. Sleep makes people vulnerable before the predators.

Reference translation

睡 眠

许多科学家一直在思考为什么睡眠对我们的大脑有如此大的促进作用。毕竟，如果我们根本不需要睡觉，那将是非常理想的：睡觉会使动物容易受到捕食者的攻击。他们认为有两个主要原因使睡眠很重要：睡眠帮助我们修复和恢复我们的器官系统，包括我们的肌肉、免疫系统和各种其他激素。睡眠对记忆起着至关重要的作用，帮助我们记住在工作或学校里学到的东西，以备将来使用。

科学家发现，适当的睡眠似乎有助于我们的免疫系统发挥最佳功能。当我们的身体处于休息状态时，被称为T细胞的免疫细胞在我们的身体里奔跑。其他的免疫细胞在睡眠充足的情况下也工作得更好。研究人员研究了人的身体对疫苗——靶向免疫系统的药物——经过一整夜的休息和完全没有睡眠之后的反应。他们发现，在疫苗接种后第二天晚上获得适当的睡眠，疫苗对病毒能产生更强烈的免疫反应。

Passage 54　　An Anti-aging Method

The best anti-ageing technique could be standing up, scientists believe, after discovering that spending more time on two feet protects DNA. A study found that too much sitting down shortens telomeres, the protective caps which sit at the end of chromosomes. People who were frequently on their feet had longer telomeres which were keeping the genetic code safe from wear and tear. Spending less time on the sofa but more time standing on both feet could help people live longer by preventing their DNA from ageing. Intriguingly taking part in more exercise did not seem to have an impact on telomere length. There is growing concern that sitting and sedentary behavior is an important and new health hazard of our time. A reduction in sitting hours is of greater importance than an increase in exercise time for elderly risk individuals. Telomeres stop chromosomes from fraying, clumping together and "scrambling" genetic code. Scientists liken their function to the plastic tips on the ends of shoelaces, and say that lifespan is linked to their length.

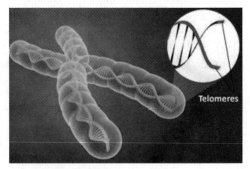

Word bank

1. telomere ['teləmɪə] *n.* 端粒（在染色体端位上的着丝点）
2. chromosome ['krəuməsəum] *n.* 染色体
3. intriguingly [in'triːgiŋli] *adv.* 有魅力地；有趣地
4. sedentary ['sedntri] *adj.* (工作、活动等) 需要久坐的；(人) 惯于久坐不动的
5. hazard ['hæzəd] *n.* 危险　*v.* 冒险；使遭受危险
6. reduction [rɪ'dʌkʃn] *n.* 缩减
7. fray [freɪ] *v.* 磨损；磨破
8. clump [klʌmp] *v.* ~ together 聚集；被聚集成群
9. scramble ['skræmbl] *v.* 行走；攀爬；扰乱

Exercises

I. Fill in the blanks with the words given below. Change the form where necessary.

| intriguingly | fray | sedentary | hazard |
| clump | scramble | telomere | reduction |

1. The _____ of unemployment should be paramount.
2. He became increasingly _____ in later life.
3. These discoveries raise _____ questions.
4. This material _____ easily.
5. Everybody is aware of the _____ of smoking.
6. Galaxies tend to _____ together in clusters.

II. Comprehension of the passage.

Choose the best answer to each of the following questions.

1. Telomeres shortening for human is _____.
   A. rare
   B. beneficial
   C. harmful
   D. insignificant

2. Which of the following is true according to the passage? _____
   A. Lifespan has nothing to do with telomere.
   B. More exercise can make telomeres longer.
   C. Less sitting can prevent DNA from aging.
   D. Telomeres make up of chromosomes.

3. Sedentary behavior may not cause _____.
   A. fatness
   B. premature ageing
   C. disease
   D. early death

4. In order to prevent aging, we should _____.
   A. exercise more   B. do sauna
   C. enjoy a hot spring   D. stand up

· 163 ·

Reference translation

一种抗衰老的方法

科学家们发现站立时间越久对 DNA 越有益后认为,最好的抗衰老的办法可能就是站起来。一项研究发现,久坐会缩短端粒,即位于染色体末端的保护帽。经常站起来的人端粒较长,这使得遗传密码不受磨损的影响。减少坐在沙发上的时间,把更多的时间用在两脚站立上,可以防止人们的 DNA 老化,有利于人长寿。有趣的是,参加更多的锻炼似乎对端粒长度没有影响。人们越来越担心坐着和久坐的行为会对我们这个时代的健康造成新的重大危害。对于老年风险人群来说,减少坐的时间比增加锻炼时间更重要。端粒阻止染色体磨损、聚集在一起和"扰乱"遗传密码。科学家将它们的功能比作鞋带末端的塑料尖端,并表示寿命与它们的长度有关。

Passage 55　Smog Can Harm Your Heart

Smog will increase the risk of cardiovascular disease, everyone should know. However, inhaling smog will not only increase your risk of cardiovascular disease, but also affect the health of the offspring. Before the parents start pregnancy, as long as they have received smog baptism within a certain period of time, it is possible to pass this risk factor all the time.

Unlike previous correlation studies, a researcher exposed his parental mice to a 38.58 μg/m^3 PM2.5 environment 30 hours before pregnancy. It is proved that the exposure of parental mice to PM2.5 before pregnancy will seriously affect the heart function health of the offspring.

The mice participating in the experiment had 6 hours of breathing smog every day. The life of the mice breathing smog lasted for three months. When the offspring mice grew to three months old, the mice exposed to PM2.5 before pregnancy were significantly lower in weight than the mice whose parents lived in a clean environment, showing a general growth retardation.

Word bank

1. smog [smɒg] *n.* 烟雾；烟尘；雾霾
2. cardiovascular [ˌkɑːdiəʊˈvæskjələ(r)] *adj.* 心血管的
3. inhale [ɪnˈheɪl] *v.* 吸气；吸入
4. offspring [ˈɒfsprɪŋ] *n.* 后代；子孙
5. baptism [ˈbæptɪzəm] *n.* （基督教的）洗礼；严峻考验
6. correlation [ˌkɒrəˈleɪʃn] *n.* 相关性；相互关系
7. exposure [ɪkˈspəʊʒə(r)] *n.* 暴露；揭露 ~ to 面临，遭受（危险或不快）
8. retardation [ˌriːtɑːˈdeɪʃn] *n.* 延迟；迟缓；落后

Exercises

I. Fill in the blanks with the words given below. Change the form where necessary.

| offspring | baptism | smog | cardiovascular |
| retardation | inhale | correlation | exposure |

1. He was treated for the effects of _____ smoke.
2. In the process of aging, creatures showed early symptoms such as memory impairment, mental _____.
3. Jack is her only _____.
4. Smoking places you at serious risk of _____ and respiratory disease.
5. _____ and communion are two of the sacraments.
6. The sky over the city was overspread with heavy _____.

II. Comprehension of the passage.

Choose the best answer to each of the following questions.

1. What did the researcher do in his study according to this passage? _____
    A. Parental mice were exposed to a 38.58 µg/m³ PM2.5 environment 30 hours per week before pregnancy.
    B. The mice were exposed to a 38.58 µg/m³ PM2.5 environment 30 hours per week after pregnancy.
    C. The mice were exposed to a 38.58 µg/m³ PM2.5 environment 6 hours per week after pregnancy.
    D. He placed the parental mouse in the clean air.
2. What characteristics appeared when the mice grew to three months? _____
    A. They died three months later.
    B. They were lower in weight than the mice whose parents lived in a clean environment.
    C. They did not show obvious characteristics.
    D. They showed different characteristics.
3. What is the main idea of this article? _____
    A. We don't need to protect the environment.
    B. PM2.5 may affect the health of future generations.

C. PM2.5 will directly lead us to death.

D. The article describes the factors that lead to the death of mice.

4. Which of the following best describes the author's attitude towards fog and haze? _____

A. Optimistic.

B. Pessimistic.

C. Critical.

D. Worried.

Reference translation

雾霾对心脏的危害

每个人都应该知道雾霾会增加患心血管疾病的风险。吸入雾霾不仅会增加自己患心血管疾病的风险，还会影响后代的身体健康。在父母备孕开始之前，只要在一定时间内受到过雾霾的影响，就有可能将这种危险因素一直传递下去。

不同于之前的相关研究，一位研究人员将亲代小鼠在其怀孕前暴露在 38.58 μg/m³ 的 PM2.5 环境中 30 个小时，切实证明，亲代小鼠孕前暴露于 PM2.5 会严重影响子代的心脏功能健康。

参与实验的小鼠每天有 6 小时呼吸雾霾。呼吸雾霾的小鼠存活了三个月。当子代小鼠们长到三个月大时，父母孕前暴露于 PM2.5 的小鼠的体重明显要比父母生活在干净环境中的小鼠轻，表现出了整体性生长迟缓现象。

Passage 56　Two Myths of Snow

First of all, it's a myth that no two snowflakes are the same. In 1988, a scientist found two identical snow crystals that had both formed in a snow storm. Since then scientists have come to learn that snowflakes can only form into 35 different shapes. Although scientists are also unsure why exactly the various shapes of snowflake form, they have identified eight predominant shapes, with each of these eight shapes having several different variations.

Second biggest myth, snow isn't white. It's actually colorless. Snow is made up of ice particles, and ice is translucent, which means that light does not pass through it easily, but rather it gets reflected. When light hits a snowflake, it gets reflected back from the snowflake's many surfaces, often bouncing between these surfaces, and because of this that light is reflected back to our eyes as the color white.

Word bank

1. myth [mɪθ] *n.* 神话；虚构的东西；荒诞的说法
2. snowflake ['snəʊfleɪk] *n.* 雪花；雪片
3. identical [aɪ'dentɪkl] *adj.* 同一的；完全相同的　*n.* 完全相同的事物

4. crystal ['krɪstl] *n.* 晶体；水晶饰品 *adj.* 水晶的；透明的
5. predominant [prɪ'dɒmɪnənt] *adj.* 显著的；明显的；盛行的；占主导的
6. variation [ˌveəri'eɪʃn] *n.* 变化；变动；变异
7. translucent [træns'luːsnt] *adj.* 半透明的；透亮的，有光泽的
8. reflect [rɪ'flekt] *v.* 反射；反照；映出

Exercises

I. Fill in the blanks with the words given below. Change the form where necessary.

| myth | variation | identical | predominant |
| reflect | translucent | crystal | snowflake |

1. The two pictures are similar, although not _____.
2. He gave me a small white box that had _____ wrapping on it.
3. Those tine wine glasses are made of _____.
4. Contrary to popular _____, women are not worse drivers than men.
5. Currency exchange rates are always subject to _____.
6. Yellow is the _____ color this spring in the fashion world.

II. Comprehension of the passage.

Choose the best answer to each of the following questions.
1. The word "identical" in the second line probably means _____.
    A. exactly alike
    B. similar
    C. opposite
    D. different
2. People usually think the color of snow is _____.
    A. colorless
    B. callow
    C. purple
    D. white
3. The reason why people see white snow is _____.
    A. snow is white in reality

B. snow reflects light

C. snow has impurities

D. people have problems with their eyes

4. How many main shapes of snowflakes from scientists' perspective? _____

A. 8.

B. 16.

C. 35.

D. 43.

Reference translation

关于雪的两个误解

首先,"不存在两片完全相同的雪花"这一说法是错的。1988 年,一名科学家发现了两个完全相同的雪结晶。之后,科学家了解到雪花只会有 35 种不同形状的造型。虽然科学家也不清楚雪花会形成各种各样形状的准确原因,但是他们已经确定出了雪花的八种主要形状,这八种形状中的每一种又有几种不同的变化。

第二大误区是雪不是白色的,而是无色的。雪是由冰晶分子组成的,而冰是半透明的,也就是说光线不能很轻易地穿透它,而会被其反射。当光线照射在雪上,雪花的众多冰晶表面会将其反射回去,通常是在各表面间来回反射。也正因为如此,反射进我们眼睛中的光线就呈现出了白色。

Passage 57 Unmanned Aerial Vehicle

Unmanned aerial vehicle is a kind of air craft which is controlled by soldiers on ground and doesn't need any pilots on board. With this obvious merit, it has attracted much attention of armies.

The main application of drones is to spy on enemies. Many drones are capable of cruising for days without being fueled.

Also, drones can be used to fight as a fighter in war. Armies now prefer drones to fighter planes because even if a drone is shot down or crashes, it won't cause any deaths.

However, abuse of drones incurs some public protest. Some specialists worry that if drones are out of control during their duties, it may cause a disaster to civilians. What's more, many people doubt drones may violate their personal privacy when spying.

Word bank

1. unmanned [ˌʌnˈmænd] *adj.* 无（需）人操作的；无（需）人控制的；自控的
2. aerial [ˈeəriəl] *adj.* 从飞机上的；空气的；航空的；空中的
3. on board 在船（火车，飞机，汽车）上；已装船
4. application [ˌæplɪˈkeɪʃn] *n.* 适用，应用，运用；申请
5. drone [drəʊn] *n.* 无人驾驶飞机；雄峰；嗡嗡声
6. cruise [kruːz] *v.* 巡航；巡游；漫游 *n.* 巡航
7. incur [ɪnˈkɜː(r)] *v.* 遭受；招致；引起
8. civilian [səˈvɪliən] *n.* 市民；平民；百姓

Exercises

I. Fill in the blanks with the words given below. Change the form where necessary.

incur	application	on board	nonsolvent
cruise	unmanned	drone	aeria

1. We went _____ and saw these people packed shoulder to shoulder on the decks.
2. The _____ population was suffering greatly at the hands of the security forces.
3. The government had also _____ huge debts.
4. It will become the world's longest endurance _____ aerial vehicle.
5. Students learned the practical _____ of the theory they had learned in the classroom.
6. I'd love to go on a round-the-world _____.

II. Comprehension of the passage.

Choose the best answer to each of the following questions.

1. Unmanned aerial vehicles are _____.
    A. popular with military domain
    B. expensive to use
    C. potentially dangerous
    D. more efficient than fighter planes
2. It is possible that unmanned aerial vehicles _____.
    A. may replace fighter planes
    B. will be quickly applied to serve the public
    C. may be designed to spies to grab personal information
    D. can last for days without fuel
3. Manipulating drones most probably relies strongly on _____.
    A. wireless
    B. speed
    C. fuel
    D. battery
4. What is the author's attitude towards the use of drones? _____
    A. Positive.

B. Negative.
C. Objective.
D. Indifferent.

Reference translation

无人机

无人机是一种由地面士兵控制的飞行器，不需要任何飞行员在飞机上。凭借这一显而易见的优点，它引起了军队的广泛关注。

无人驾驶飞行器的主要应用是对敌人进行侦察。许多无人驾驶飞行器能够在没有燃料的情况下巡航数天。

此外，无人驾驶飞行器可以用作战争中的战斗机。相比战斗机，军队现在更喜欢无人驾驶飞机，因为即使无人驾驶飞机被击落或坠毁，也不会导致人员死亡。

然而，滥用无人驾驶飞行器引起了一些公众的抗议。一些专家担心，如果无人机在执勤期间失控，可能会给平民带来灾难。更有甚者，许多人怀疑无人驾驶飞行器在进行间谍活动时可能会侵犯他们的个人隐私。

References

[1] Science Daily. Technology[BE/OL].(2019-3-30)[2014-3-30]. https://www.sciencedaily.com.

[2] Nature. [BE/OL].(2019-3-30)[2016-11-24].http://www.nature.com/nature/ journal/v539/n7630/full/539471e.html.

[3] David Biello. Scientific American[BE/OL].(2019-3-30)[2016-8-1]. https:// www.scientificamerican.com/article/new-bionic-leaf-is-roughly-10-times-more-efficient-than-natural-photosynthesis.

[4] John Travis. [BE/OL].(2019-3-30)[2016-8-11].www.sciencemag.org/topic/robots.

[5] Daniel Bates. Beijing review [BE/OL].(2019-3-30)[2016-10-27]. http://www. nature.com/articles/srep38145.

[6] Jamie Condliffe. [BE/OL].(2019-3-30)[2015-5-29]. http://gizmodo. com/new- memory-alloy-springs-back-into-shape-even-after-10-1707636284.

[7] Dave Lee. North American technology reporter[BE/OL]. (2019-3-30) [2016-1-15]. http://www. bbc.com/news/technology-35310200.

[8] Kerri Smith. [BE/OL].(2019-3-30)[2013-6-6].https://www.nature.com/news/ dummy-jpg-7.10929?article=1.13144.

[9] Yaobinchen. [BE/OL].(2019-3-30)[2013-6-6]. http://www.sciencedirect. com/ science/book/9780123971999.

[10] Giuseppe Faraco and others. [BE/OL].(2019-3-30)[2018-1-15].https://www.nature. com/articles/s41593-017-0059-z.

[11] Robert F. Service[BE/OL].(2019-3-30)[2018-5-31]. http://www.sciencemag. org/news/2018/05/new-artificial-nerves-could-transform-prosthetics.

[12] Walking on sunshine, BBC News[BE/OL].(2019-3-30)[2018-5-31]. http://www. bbc.com/news/uk-scotland-glasgow-west-44076553.

[13] Loren E. Wold [BE/OL].(2019-3-30)[2018-12-16]. https://tech.sina.com.cn/ d/f/2018-12-16/doc-ihmutuec9633497.shtml.

[14] China Daily [BE/OL].(2019-3-30)[2018-12-11].http://language.chinadaily. com.cn/thelatest/page_3.html.

[15] Virginia[BE/OL].(2019-3-30)[2017-10-18].Morell,http://www.sciencemag.org/news/2017/10/flower-petals-have-blue-halos-attract-bees.

[16] Mitch Leslie. Science[BE/OL].(2019-3-30)[2017-11-1]. http://www.sciencemag.org/news/2017/11/researchers-find-fatal-flaw-childhood-tumors.

[17] Solids, liquids and gases [BE/OL].(2019-3-30)[2017-1-1]. https://www. sciencelearn. org.nz/resources/607-solids-liquids-and-gases.

[18] Jamie Condliffe. [BE/OL].(2019-3-30)[2015-5-29]. http://gizmodo.com/new- memory-alloy-springs-back-into-shape-even-after-10-1707636284.